IRISH FARMING

Implements and Techniques
1750 - 1900

JONATHAN BELL
and
MERVYN WATSON

Department of Material Culture
Ulster Folk and Transport Museum

JOHN DONALD PUBLISHERS LTD
EDINBURGH

ISBN 0 85976 164 9

Exclusive distribution in the United States of
America and Canada by Humanrites Press
Inc., Atlantic Highlands, N.J. 07716, USA.

Phototypeset by Newtext Composition Ltd., Glasgow.
Printed in Great Britain by Bell & Bain Ltd., Glasgow.

Preface

This book attempts to describe some of the major changes in Irish cultivation techniques since 1750. We have not tried to produce an exhaustive inventory of every implement used on Irish farms during the period, but have attempted to show links between the techniques in themselves and also, at least in rudimentary outline, how these connect to wider social and economic changes. The subject is so big that the gaps in our work will be obvious to many readers. In dealing with topics such as land reclamation, we were overwhelmed by the amount of published material to be examined. Researching other topics, however, such as the use of *graffáin* in Munster, or 'making' land on the Aran islands, was frustrating because we knew that we were only skimming the surface of complex techniques, as detailed data were not easily available. The book, therefore, should be read as an introduction, but we hope that it may convince some other people that more detailed research would be worthwhile.

Much of the material presented has been collected as part of our work in the Ulster Folk and Transport Museum, but we have also relied heavily on other people's research. The work of R.A. Gailey, C. Ó Danachair, and A.T. Lucas has been of particular importance. Most of our knowledge of farming, however, we owe to the many Irish farmers whom we have met during fieldwork. Their patience with the limited understanding of two Belfastmen has been universal, and we can only hope that our ignorance amused rather than annoyed them.

We have some other people to thank. Patricia Stratford rescued us with high technology, while the photographic staff of the Ulster Folk and Transport Museum were their usual efficient selves. We are also grateful to institutions from which we have received permission to reproduce illustrations. These include the National Museum of Ireland, An Clódhanna Teoranta, the *Irish Times*, and the Committee for Aerial Photography of Cambridge University.

Contents

Tory

Bloody Foreland
Gweedore

Carndonagh

Coleraine

Templemoyle

ANTRIM

DONEGAL
Glenfin

DERRY

Newtownstewart

TYRONE

Belfast

Bangor
Comber

Ederney

ULSTER

Moira

Erris Head

Lisadell

Enniskillen

FERMANAGH

ARMAGH

DOWN

Ballina

SLIGO

Clones
MONAGHAN

Ballybay

Newry

Mourne Mtns

Achill

MAYO

Castlebar

LEITRIM

CAVAN

Bailieboro

LOUTH
Ardee

Westport
Murrisk

ROSCOMMON

LONGFORD

Inisbofin

Packenham

WESTMEATH

Ardbraccan

CONNAUGHT

MEATH

Athlone

Cloncurry

DUBLIN
Cellbridge

GALWAY

Galway

Ballinasloe

LEINSTER

Tallaght

Aran Islands

Aughnish

OFFALY

KILDARE

CLARE

Ennis

LAIOS

WICKLOW

TIPPERARY

CARLOW

Limerick

Adare

KILKENNY

Cashel

WEXFORD

LIMERICK

Cullen
Tipperary

Arbella

Clogheen

Wexford

Dingle

MUNSTER

WATERFORD

Forth

Killarney

CORK

Fermoy

KERRY

Nedeen

Imokilly
Youghal

Cork

Clonakilty

0   50 mls

CHAPTER 1

Landlords, Farming Societies and Agricultural Improvement

Very few people would be surprised to learn that Irish cultivation techniques have changed greatly during the last two and a half centuries. Parallel changes have been recorded throughout Europe, America and, more recently, most other parts of the world. What has intrigued observers in the past, however, and still concerns historians, is that older practices have been particularly slow to die out in Ireland. In fact, it often seems that new techniques have simply been added to the already existing range of available methods of cultivation, rather than replacing them. Even today it is still fairly easy to find practices which would have been familiar to farmers in the eighteenth century.

Many different sorts of argument have been put forward by agriculturalists, historians and others to explain this persistence or 'lag'. A large section of Irish historical writing is concerned with the structures of society which provide the basis on which any understanding of farming practices must be built. These include state institutions, systems of landholding, and patterns of trade. Labour, the supply of capital, and the implications of population changes, have also been discussed at length. However, although historians have suggested ways in which these aspects of society have influenced farming methods, this has not often been a central theme in their work. This book will have a different emphasis. In it, we want to concentrate on farming techniques in themselves, both individually and as elements in particular systems of farming. This will sometimes mean referring to the wider topics just listed, and indeed one of our aims is to show that a detailed examination of implements and techniques is relevant to these. However, in general it will be the mechanics of crop production which will be emphasised.

Our central aim in the book is to show the rationality of the farming practices we describe. In doing this we also hope to show how unsatisfactory some other explanations of the practices have been. Several of these are as old and as remarkably persistent as some of the practices examined. Assertions which we find particularly unhelpful are that Irish farmers were simply prejudiced against change, or were

1

'naturally conservative', and that in part this arose from laziness, or to use the more discreet modern phrase, from a 'high leisure preference'.

In collecting material to use in the book, we have made use of a wide variety of sources, both documentary and oral. These will be cited where relevant. For the eighteenth and nineteenth centuries, documentary sources were obviously most important. The character of these determined not only the kind of information available, but also throws light on how the knowledge of innovations in farming was disseminated, and by whom. The number of works on Irish farming multiplied after 1750. Most of these were intended to lead to the improvement of cultivation methods, and common practices were usually only described to be condemned. Before the late nineteenth century, the authors of these works were mostly either English, or members of the Anglo-Irish ascendancy. These men sometimes attempted to write especially for Irish small farmers, but in general their readership consisted of Irish landlords and gentlemen farmers. As we shall see, the farming methods which were available to these men were often neither within the resources of the mass of Irish farmers, nor were they always the most suitable for their situation.

Landlords and Tenants

Until the late nineteenth century, most farmers in Ireland were tenants on large estates. This had important implications for their ability, and even desire, to make radical changes in their farming methods. The implications were complex, however, and there has been strong disagreement between both contemporaries and modern historians as to what they were. Tenancy meant that some farm produce had usually to be set aside for the payment of rent, and therefore less could be accumulated for investment. However, it has also been claimed that high rents made tenant farmers work harder, and if smallholders were forced off the land because of their inability to pay, that was not necessarily a bad thing economically, since it produced holdings of a more viable size, and also more commercially orientated farmers. (From this point of view the fate of the people forced off their farms is regrettable, but basically irrelevant.)

Leasing arrangements on farms have also been accepted as having important influences on farming methods, but again there is disagreement as to their real effects. Some historians have argued that long leases hindered improvement, since landlords could not organise large-scale schemes while their lands were occupied by small farmers with security of tenure. Contemporary writings, on the other hand, contain many emphatic claims that short leases, or no lease, discouraged the tenants from engaging in anything other than short-term improvements. The complexity of the question of leases is well illustrated by an example from county Roscommon in 1832. The situation described was one where small farmers were living on the edge of a bog, letting their cattle on to it for common pasturing:

> These farmers have usually terms of lives or years in their holdings, too short to tempt them, even if possessed of capital and skill, to enter on the permanent improvement of the bog, while they are yet abundantly sufficient to render such an operation impracticable for the landlord. The landlord has demised to the tenant a vague possession of what he considered of little or no value; the shortness of the lease obliges the tenant to leave his holding in its unprofitable state, but were the landlord to propose to improve it, the tenant having got a present right to prevent him, that right would become valuable, just in proportion to the extended exertions of the landlord, and would inevitably be set up by the tenants.[1]

One consequence of the system of landholding which had immediate implications for the techniques used in the cultivation of even a single year's crop, was that farms of very different sizes were produced. In the late eighteenth century, Arthur Young visited a farm belonging to Mr McCarthy of Springhouse, Tipperary. Young believed that this farm of 9,000 acres 'must be the most considerable one in the world'.[2] There were also many large and medium-sized farms owned by lesser gentry or rented by wealthy tenants. However, especially before the famine of 1845, there was also a vast number of tiny holdings. These could be produced by subdivision, where farmers attempted to provide all their sons with land, or by the system of rundale, where several families would lease an area of ground and divide it into strips which were periodically redistributed between participant farmers. In Ulster, small farms were particularly common, their survival helped in the late eighteenth and early nineteenth centuries by the linen industry, which allowed the development of a dual economy, families farming part-time, but also working at aspects of linen production. At the extreme end of the scale, cottars attempted to provide a subsistance living for themselves and their families on holdings of an acre or less. Confronted with this grim struggle for survival, it is not surprising that agriculturalists interested in large-scale, and often risky, changes concentrated their attentions on the wealthy.

Just how eager large Irish farmers and landlords were to introduce new methods was, and is, a matter of controversy, however. As with so many Irish historical questions, general statements must be based largely on impressionistic evidence. In general, contemporaries did not think that the wealthy were doing all they could. Arthur Young was vitriolic in his condemnations of minor landlords and middlemen, who lived off rents obtained by sub-letting land:

> They [middlemen] are . . . sometimes resident on a part of the land they hire, where it is natural to suppose they would work some improvements; it is however very rarely the case. I have in different parts of the Kingdom seen farms . . . in which the residence of the principal tenant was not to be distinguished from the cottared fields surrounding it. I was at first much surprised at this but after repeated observation, I found these men very generally the masters of packs of wretched hounds, with which they wasted their time and money, and it is a notorious fact that they are the hardest drinkers in Ireland.[3]

In 1824, Hely Dutton claimed that in county Galway the attitudes of many landed proprietors towards their tenants was summed up in the statement, 'I do not care a damn what they do, or how dirty their houses are, so I receive my rents'.[4]

These general condemnations must of course be treated with caution, because they are based on impressions rather than systematic analysis, and also because they were made by men who believed that agricultural innovation was of overwhelming importance and who were therefore critical of anything other than total dedication to it. It should also be noted, however, that they were not the opinions of nationalists or social radicals, but of men who in most other respects would ally themselves to the interests of landlords and gentlemen farmers. It is significant also, that even among modern historians who wish to debunk the 'traditional' view of Irish landlords, which they see as 'strongly biased in spite of its plausible coherence and sustained longevity',[5] it is acknowledged that landlords in general did not invest large sums on their estates.'A gate here, a fence there, a short stretch of road or a pedigree bull bought in a fit of enthusiasm and then sold when his appetite for oil-cake became burdensome, was the usual erratic pattern of landlord investment in Ireland.'[6] However, despite the fact that they were probably untypical, a lot of this book will be taken up with descriptions of innovations introduced by wealthy farmers. This active minority does have a central importance, since it was through farmers of this kind that knowledge of many new techniques came into Ireland, and it was also from these people that members of farming societies and other bodies were drawn, which had the explicit aim of improving Irish farming methods. The rest of this chapter will outline the history of some of these organisations and attempt to assess how far they were effective in spreading the gospel of improvement.

Agricultural Societies and Schools.

In 1844, the Rev. William Hickey ('Martin Doyle') argued that agricultural societies, schools and periodicals were crucial to the development of farming in Ireland, 'for in that country there is not, as in England and Scotland, a class of farmers capable of teaching, in their everyday operations the principles and practice of improved husbandry'.[7] The most famous, and earliest, body to attempt to fill this gap, the Dublin Society, was founded in 1731 by a small group of gentry, with the aim of improving not only the system of farming, but also the manufactures and 'useful arts and sciences' in Ireland.[8] Arthur Young hailed the society as 'the father of all the similar societies now existing in Europe'.[9] From its earliest days the society attempted to increase awareness of new farming techniques, mostly those developed in England. One of the first projects undertaken was to publish Jethro Tull's *Horse hoeing husbandry* in Dublin. Others included sponsoring of experiments in methods of cleansing corn and clover grass seed, and holding trials of new ploughs, first organised in Phoenix Park, Dublin, during the winter of 1732.[10] In 1733, a display of new farm implements was set up in a vault of the Irish parliament building, and in 1741 a system of premiums was begun, these first being paid to encourage the careful cultivation of wheat, barley and turnips, and for manuring practices and the successful cultivation of fruit trees. From its earliest days, the society did attempt to reach others besides gentlemen. In 1737, its policy was officially stated to be the instruction of 'the poorer sort, husbandmen and manufacturers'.[11]

The emphasis of the society on establishing English standards of farming led to the employment of several Englishmen as instructors. In 1749, it was announced that a Mr John Cam, who was claimed not only to be an efficient farmer, but also to be skilful in making ploughs and carts, had been engaged

> as an itinerant husbandman, to advise [farmers] . . . in the right way of ploughing their land for the growth of corn. He will carry with him some ploughs of his own making, etc. Said Cam will set out from Dublin on Monday, 27th, and will go to Navan, and so proceed to the rest of Co.

Meath, and the Counties of Kildare, Carlow, Kilkenny, etc., where he
may meet growers of corn, and instruct them in the right way of tillage, and
thereby save labour, expense and time.[12]

In 1763, the Englishman, John Wynn Baker, after writing a pamphlet
for the Dublin Society, established a farm at Loughlinstown, near
Cellbridge, county Kildare. Baker, assisted by the society, took five boys
from the Foundling Hospital, and began to instruct them in making
farm implements. The Society began to pay him a salary in 1769, and in
1771 he also received a grant to open a factory.[13] After his death a few
years later, his work, and the patronage of the Dublin Society for it, were
praised by Arthur Young. 'Nothing was ever better imagined, than the
plan of fixing an English farmer in the Kingdom, so much at the
Society's expense, as to give them a power over part of his management
. . . it was also a very wise measure to enable him to establish a
manufactory of husbandry implements.'[14] Young also criticised the
Society, however, first for not supporting Baker's work more liberally,
and also for allowing him to set up his farm on land which was already
well prepared for cultivation. Young argued that it would have been
better to have settled him on a mountainous tract of land which included
bog as well as workable soil. Related to this, Young believed that Baker
should have been encouraged to develop successful practical schemes,
rather than to engage in experiments which might be 'objects of curiosity
to a private specialist, but quite unworthy of the Dublin Society'.[15] Baker
was the best-known of the Englishmen encouraged by the Society to set
an example to Irish farmers. In 1780, a Thomas Dawson was employed
as an instructor, but he seems to have been an itinerant instructor and
did not develop the ambitious projects undertaken by Baker.

In 1781, the Society reopened a 'museum' in which to display new
agricultural implements, this time in Poolbeg Street, Dublin. This was
later extended and became a factory where experimental implements
could be purchased. For most of the eighteenth century, the Society was
given a grant by the Irish parliament, but after the Act of Union in 1800,
this was systematically reduced. In 1801, however, the Society
undertook one of its most important projects: the compilation of
'statistical surveys' of individual counties in Ireland. By 1831, surveys of
twenty-three counties had been published.

In 1820, King George IV became the Society's patron and the prefix
'Royal' was added to its name. During this period also, however, several
other related national organisations were formed, which concentrated
exclusively on agricultural improvement. A Farming Society of Ireland
was founded in 1800, and after its incorporation in 1815 (Fig. 1) was

Fig. 1. A medal of the Farming Society of Ireland, showing an improved swing plough pulled by two oxen, all controlled by one man (UFTM, L2318/2).

given a grant of £5,000 per annum. This society was short-lived, however, and its grant was stopped in 1828. An agricultural Improvement Society, established in 1841, was more successful. It was incorporated as the Royal Agricultural Society of Ireland in 1860, and functioned for forty years.[16] The Society hoped to organise an annual show in one of the four provinces, to re-open an agricultural museum, and to establish an agricultural college.[17] It also, along with the Royal Dublin Society (into which it was absorbed in 1887), assisted local farming societies, until the establishment of the Department of Agriculture and Technical Instruction in 1900.

The Royal Dublin Society seems to have been a creative, energetic body, especially when we consider its major achievements, such as the County Statistical Surveys or the Irish Industrial Exhibition of 1853. However, contemporaries were by no means united in praise of the organisation. In the early nineteenth century, for example, the society was said to be looked on 'with a feeling not of actual suspicion but of distrust, as a body which does nothing'.[18] The comment of another observer was hardly more complimentary. 'I believe it to be an inefficient society, but not a jobbing society.'[19] These comments were made at a time when Irish agriculture was depressed, and by the 1840s

Martin Doyle saw evidence of change: 'The Royal Dublin Society . . .
for a long period lost sight of the principal objects of its association . . .
but within the last few years has again applied a part of its resources to
the promotion of agriculture'.[20] A British text of the same period, while
seeing the Society as giving only 'a slight impulse' to Irish agriculture,
claimed that this was not through the fault of its members, but was due
to the situation in which it operated, where farmers were ignorant and
had insufficient capital, and landlords were indifferent to the condition
of their tenants.[21] It has also been argued that the depression which may
have impeded the work of the Royal Dublin Society led to the actual
demise of the Farming Society of Ireland in 1828.[22] The Royal Dublin
Society survived and has continued to function until the present. One
important scheme started in 1891, and based around Swinford, county
Mayo, was aimed at encouraging small farmers to adopt new methods.
An instructor, Mr D. O'Dowd, organised the cultivation of crops on
selected holdings. These were intended to act as an example to
neighbouring farmers. The most important results of the project were in
the testing of potato varieties and sprays. These will be described in a
later chapter.

The development of national societies was paralleled by the growth of
local farming societies which were also being established from the mid-
eighteenth century onwards. The earliest were in Antrim, Kildare and
Louth. Later in the century societies were also formed in counties Mayo,
Roscommon, Fermanagh and Clare.[23] Societies multiplied during the
nineteenth century but varied a lot in size and in the range of their
activities. Many held ploughing competitions and organised local shows,
but several also attempted to set up agricultural schools.

These societies were praised and criticised as much as the Royal
Dublin Society. The resolutions passed at a joint meeting of the Cork
Institution and Cork Farming Society in 1807 show how ambitious their
aims could be. These included the distribution of good seed, the
encouragement of the reclamation of waste land, spreading knowledge of
improved implements, and extended crop rotations. Premiums were to
be given as an incentive.[24] However, in 1834, while praising the
effectiveness of some societies which did not 'soar too high', but
concentrated on practical schemes with obvious advantages to small
farmers, a contributor to the *Irish farmer's and gardener's magazine* was
sharply critical of the performance of societies in general:

> That the agricultural societies are not extending their usefulness in a way
> which might be desirable, will appear from the fact that in districts

containing a population of probably, a thousand, holding farms of greater or lesser extent, we see a farming society consisting of not more than ten individuals. What a ridiculous exhibition for such a society to be employed time after time making exclusive rules, and in all the pride of an assumed self-importance, endeavouring to withhold any little knowledge they may possess from those who stand most in need of such, for to this and no less, amount a great number of their regulations. We would just ask what has the exhibition of a few overfed animals to do with the agricultural interests of the country?[25]

Even Martin Doyle was sometimes discouraged by the apparent lack of success of farming societies. Discussing the attempts of two Wexford associations to introduce green cropping, he commented:

they have laboured hard at this point with little success, the cultivation of green crops being confined either to the higher and wealthier class of farmers (who would grow them without the stimulus of a premium), or else to small holders coaxed to cultivate them in small patches for a short time, while the reward was certain, and reverting to the old system on the cessation of the bribe.[26]

Differences in opinion as to the effectiveness of societies can also be found in testimonies given to the Devon Commission during the 1840s. One small farmer, Laurence Egan, recorded a hymn of praise to the Ballinasloe Agricultural Society. Before the formation of this society, Egan said that he was a farmer in name only. 'I may therefore, date the commencement of my farming with the commencement of the . . . Society . . . Not that I am yet an experienced farmer; far from it, for I have much yet to learn. However . . . I am a child of the society's.'[27] In contrast, however, John Brennan, a land agent from county Kerry, was abrupt: 'The agricultural societies do not afford much benefit to the small farmer. They do not know how to expend money if they had it'.[28]

Martin Doyle believed that the major success of farming societies was in the introduction of new ploughs and ploughing techniques, and this was also the conclusion of the Cork Farming Society when it reviewed its own activities.[29] Doyle, however, also believed that model farms and agricultural schools were by far the most efficient way to illustrate the advantages of innovation. He worked along with a local Wexford society to set up a school at Bannow. The school opened in 1821, but lasted only seven years. Doyle attributed its failure to a lack of funds, and also because he, a Church of Ireland cleric, was moved to another parish.[30] In 1847, the Devon Commission published details of the rules and timetables of six agricultural schools, each in a separate county.[31] Of these the most famous were at Templemoyle in county Derry, and Glasnevin in county Dublin.

Plans to establish a seminary at Templemoyle were drawn up at a meeting of the North-West of Ireland Society in Derry in 1823. 'It was considered by several members of the Society that . . . good would arise to the public, if . . . a school were established, in which the sons of farmers should . . . be instructed in the *scientific* and *practical knowledge* of farming'[32]

Templemoyle was not primarily intended for the sons of gentry, who were usually sent to school in Britain, but to turn out working farmers. For this reason, it was stated in the rules that at any time one half of the pupils would be employed in manual labour, although while they were working the head farmer would explain the reasons for using the particular technique. The other half of the pupils' time was to be spent in 'literary' work.[33]

The Templemoyle school was privately financed until 1850, and then continued under the National Board of Education until its closure in 1866. By this time, however, the National Board had opened several Model Agricultural Schools, the earliest of which was to instruct teachers being trained in Dublin 'in the principles and practice of improved agriculture'.[34] In 1862, a model farm of five and a half acres was started at Glasnevin, tilled using spade labour.[35] Practical farming was also included in the curriculum of many ordinary national schools. In the most common arrangement, the local teacher was encouraged to rent a small piece of ground which he used as an 'illustrative model of scientific principles'. Pupils worked on these farms by choice, and were paid for doing so.[36] Until the 1860s, small areas of ground attached to workhouses were also cultivated, but despite the protests of agriculturalists, the grants given for these were withdrawn in 1863. At the same time financial cuts were also made at Glasnevin, four lecturers' posts being cancelled, and the intake of students reduced by twenty.[37]

If we examine comments made on the performance of the Templemoyle School, they are as divided between praise and criticism as those on the farming societies.

One of the most celebrated visitors to Templemoyle was William Makepeace Thackeray, who praised the scenes of intelligent, hard work which he recorded there, comparing them to what he presented as the degenerate lifestyle created at English public schools such as Eton (Fig. 2). Templemoyle was different, he claimed, because what it taught was useful.[38] He found the buildings at Templemoyle spacious, simple and comfortable, the pupils models of decorum, and the balance between

Fig. 2. Work in a potato field at Templemoyle agricultural school, as observed and praised by W. M. Thackeray. The spade shown, however, is a one-sided loy, a type often criticised by improvers (Thackeray, W. M. (1879), p.325).

literary and practical work well adjusted: 'At certain periods of the year, when all hands are required, such as harvest, etc., the literary labours of the scholars are stopped and they are all in the field . . . In winter the farm works do not occupy the pupils so much, and they give more time to their literary studies'.[39] Thackeray acknowledged that there had been some early difficulties in setting up the school, but at the time of his visit found that the £10 in fees a year charged to each pupil, along with the produce of the farm, made it self-supporting.[40]

Even if we accept Thackeray's account, however, it seems that he may have visited the school during a very brief golden age. Edmund Murphy visited Templemoyle in 1834, and found that although there was accommodation for one hundred pupils, only thirty-five were boarding in the school. Murphy also examined the agricultural work, and although he inspected excellent examples of fencing, draining and clearing stones, he found no facilities for the winter housing of cattle, no piggery, no sheepfold (or sheep) and no systematic storing of manure. 'We expected that at least on a small scale, experiments would be going on . . . but nothing of the kind has been attempted.'[41] Commenting on the small number of pupils, Murphy concluded that what he had found at Templemoyle 'gave sufficient reason for this apathy'.[42]

Murphy had some praise for the literary elements in the Templemoyle curriculum, but this was not the conclusion of inspectors from the National Board of Education, shortly after the school was taken over in 1850. In 1856, for example, an inspector found that the standard of answering in written examinations was 'in general very bad'.[43] In 1870, several years after the closure of Templemoyle, the Powis Commission produced a negative assessment of Model Schools of Agriculture in general. They did not justify the money spent on them.[44] Contemporaries and modern historians have pointed out that the schools were not only regarded with apathy by small farmers, but with suspicion. The emphasis on practical skills was intended to prepare pupils for their expected station in life, but also designed to prevent them from wishing to rise above it. An official account of Templemoyle stated that the school would 'elevate the intellectual and moral condition of the pupils, without incapacitating them for their ordinary occupations'.[45] When it is recognised that improvers often attempted to persuade landlords to improve their estates, by pointing out that an affluent tenantry would be better able to pay increased rents, it is understandable that small farmers suspected that improvers, landlords and agricultural schools were all primarily concerned to make them able to pay more money. This was made explicit by a National Board Inspector, reporting in 1851: 'Farmers and peasantry . . . regard [agricultural schools] . . . as agencies instituted and supported for the purpose of keeping up high rents'.[46]

Later attempts to organise societies and other bodies, at least partly concerned with agricultural education, included the Congested Districts Board, established in 1891, and the central body of the co-operative movement, the Irish Agricultural Organisation Society, founded in 1894. The Department of Agriculture and Technical Instruction, founded in 1900, has already been mentioned. These organisations were grant-aided by the state, and their intentions, and effectiveness, have been as energetically debated as any of the earlier bodies discussed in this chapter. Whatever conclusions we draw from these debates, however, the organisations have importance for the contents of this book. First, the large volume of reports and periodicals they commissioned, along with nineteenth-century farming journals, provide us with our main source of data on farming practices in Ireland before the beginning of this century. Second, because they were concerned with innovation, they can at least show us, if not common practice, at least which methods were regarded as most advanced at any period. The conclusion of this book will argue against the rigid views they often expressed, and defend some practices which they, in some cases, contemptuously dismissed. The book would not have been possible, however, without the records they have left.

NOTES

1. Weld, Isaac, *Statistical survey of the county of Roscommon* (Dublin, 1832), Appendix, p.lii.

2. Young, Arthur, *A tour in Ireland*, vol. 2 (Dublin, 1780), p.457.

3. *Ibid.*, 'General observations', p.18.

4. Dutton, Hely, *A statistical and agricultural survey of the county of Galway* (Dublin, 1824), p.70.

5. Vaughan, W.E., 'Landlord and tenant relations in Ireland between the famine and the land war, 1850-78', in *Comparative aspects of Scottish and Irish economic and social history*, eds. L.M. Cullen and T.C. Smout (Edinburgh 1978), p.223.

6. *Ibid.*, p.22.

7. Doyle, Martin, *A cyclopaedia of practical husbandry*, rev. ed. W. Rham (London 1844), p.16; Doyle Martin, 'On Agricultural schools', The Irish farmer's and gardener's magazine, vol. 1 (Dublin, 1834), p. 313.

8. Berry, Henry F., *A history of the Royal Dublin Society* (London, 1915), pp.5-6.

9. Young, Arthur, *op. cit.*, vol. 2, 'General observations', p.95.

10. Berry, Henry F., *op. cit.*, pp.18-21.

11. *Ibid.*, p35.

12. *Ibid.*, p.50

13. *Ibid.*, pp.137-138.

14. Young, Arthur *op.cit.*, vol. 2, 'General observations', p.95.

15. *Ibid.*, p.101.

16. Curran, Simon, 'The Society's role in agriculture since 1800;' in *The Royal Dublin Society, 1731-1981*, eds. J. Meenan and D. Clarke (Dublin, 1981), pp. 88-91.

17. Berry, Henry F., *op. cit.*, p.297.

18. Meenan, J. and Clarke, D., 'The Royal Dublin Society', in *The Royal Dublin Society, 1731-1981*, eds, J. Meenan and D. Clarke (Dublin, 1981), p.30.

19. *Ibid.*

20. Doyle, Martin, *op. cit.* (1844), pp.20-21.

21. Library of Useful Knowledge, *British husbandry*, vol. 1 (London 1834), p.30.

22. Curran, Simon, *op. cit.*, p.90.

23. Berry, Henry F., *op. cit.*, p.222.

24. *The Munster farmer's magazine*, vol. 1 (Cork, 1812), pp.11-12.

25. 'F.', 'Observations on the agriculture of the north of Ireland', *The Irish farmer's and gardener's magazine*, vol. 1 (Dublin, 1834), pp.122-123.

26. Doyle, Martin, *op. cit.*, (1834), p.313.

27. [Devon Commission], *Digest of evidence taken before Her Majesty's Commissioners of Inquiry into the state of the law and practice in respect to the occupation of land in Ireland*, part 1 (Dublin, 1841), p.29.

28. *Ibid.*

29. *The Munster farmer's magazine*, vol. 4 (Cork, 1816), p.78.

30. Doyle, Martin, 'On agricultural schools', *The Irish farmer's and gardener's magazine*, vol. 1 (Dublin 1834), p.551.

31. [Devon Commission], *op. cit.*, pp.35-61.

32. Murphy, Edmund, 'Observations made on a visit to the agricultural seminary at Templemoyle, in the county of Londonderry', *The Irish farmer's and gardener's magazine*, vol. 1 (Dublin, 1834), p.486.

33. Kennedy, David, 'Templemoyle agricultural seminary, 1827-1866', *Studies: an Irish quarterly review*, vol. 29 (Dublin, 1940), pp.122-123.

34. Dennchy, Mary, 'Agricultural education in the nineteenth century', *Retrospect: Journal of the Irish history students' association*, vol. 5 (Dublin, 1982), p.50.

35. Baldwin, Thomas, *Introduction to Irish farming* (London, 1874), p.114.

36. Dennchy, Mary, *op. cit.*, p.51.

37. *The Irish farmer's gazette*, vol. 22 (Dublin, 1863), p.191.

38. Thackeray, William Makepeace, *The Irish sketch-book of 1842 (Works*, vol. 18) (London, 1879), p.330.

39. *Ibid.*, p.329.

40. *Ibid.*, pp.326-327

41. Murphy, Edmund, *op. cit.*, p.490.

42. *Ibid.*

43. Kennedy, David, *op. cit.*, p.123.

44. Dennchy, Mary, *op. cit.*, p.51.

45. Kennedy, David, *op. cit.*, p.122.

46. *Ibid.*, p.125.

CHAPTER 2

The Preparation of Land for Crops

The different stages involved in preparing land for cultivation can be identified most easily by examining schemes for improving marginal mountain land and bogs. These schemes became increasingly common in Ireland after the mid-eighteenth century, and involved some of the richest landlords and their poorest tenants. Many of the differences in the techniques used can be linked to the resources available to particular farmers. All land 'reclamation' required intensive labour, however, and some techniques were used on large and small schemes alike. One method used by the smallest farmers, but also found in many large-scale projects, was the use of high, narrow cultivation ridges. The importance of these ridges, not only in reclamation, but in Irish sytems of cultivation generally, will be discussed later in this chapter.

The Devon Commission's report, published in 1847, recognised three main approaches taken to land reclamation by Irish landlords:

1. The landlord could organise the entire reclamation scheme, and only after it was complete, let the land to tenants.

2. Tenants could be allowed to settle on marginal land, to reclaim it, with some assistance from the landlord. This assistance might include monetary grants or loans, training in approved techniques, or might be confined to the inducement of several years' tenancy at a lower rent.

3. The landlord could leave the whole operation to the unassisted labour of his tenants. Most contemporaries, and modern historians, agree that this last approach was the one most commonly taken by landlords, until the mid-nineteenth century.[1]

By the late eighteenth century, there were some very large-scale reclamation schemes organised by landowners. Arthur Young, for example, gave the title 'the Great Improver' to Lord Chief Baron Forster of Cullen, after inspecting a scheme which brought 5,000 acres into cultivation. This land, which had previously been 'a waste sheep walk, covered chiefly with heath and some dwarf furze and fern', had been transformed into 'a sheet of corn'. The improvement had required the erection of twenty-seven lime kilns, and the construction of hollow,

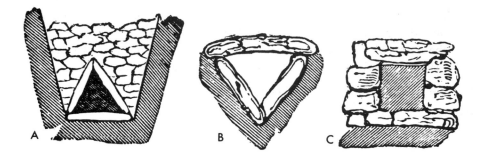

Fig. 3. Three methods of constructing stone drains. Agriculturalists warned that A and B were particularly liable to collapse (Doyle, M. (1844), p.204).

underground drains (Fig. 3) required 'the burying in . . . of several millions of loads of stones'.[2]

Some landowners who organised major reclamation schemes, first had roads built through the area to be improved. Road building and the drainage of land were seen as interconnected. In 1802, one writer pointed out that the effective drainage of mountains and bogs could not even be attempted if access was impossible, while road-building, on the other hand, could not be carried out on very wet, undrained land.[3] The best-planned reclamation schemes involved the co-ordinated construction of both drains and roads.

Drainage

Some drainage was organised on a large scale during the later eighteenth century. Arthur Young recorded several examples of rivers being widened, or navigable channels cut through bogs. One spectacular scheme was undertaken by Robert French of Moniva Castle, county Galway. Before drainage, the castle was surrounded by 'two very high, red, deep wet bogs, impassable for any beast of burden, very difficult for even men to pass'. The bog on the east side of the castle had grown up so high that it blocked the view from ground-floor windows. A river between this bog and the castle overflowed during heavy rains. There were some small drains already cut through the bog, but in 1774 French had them widened and deepened:

> The sides of the drain were so high, that I was obliged to [have them] cut . . . in some parts into benches, in the form of stairs, to prevent the men

at the bottom from being overwhelmed, which would once have happened, only that a man standing on the surface, observing the bog to burst, gave the alarm, by which he saved the lives of several men; for in a few moments many perches in length of the drain were filled up to the top.

French had the river flowing near the castle diverted, and so produced ten acres of meadow. The new river course had a fall which was used to power a bleaching mill. Another drain cut from a nearby lake was used to power the mill in summer. Some of the wooden piles used to hold the new river banks in place were washed away during heavy rain, but these were replaced by stones carried by boat. Boats were also used on the broadest drains cut through the bog. One drain was cut so that boats could carry manure right in to the castle's farm yard. The boats also carried limestone and manure around the bog, to points from which they were spread as fertiliser. French claimed that when the scheme was completed, not only was the land greatly improved, the bog had subsided fifteen to twenty feet, and no longer blocked the view from the castle's windows.[4]

Young recorded the construction of underground 'Hollow' or 'French' drains, not only under the direction of great landowners, but by some smaller farmers as well. At Kilrue, for example, he found that 'Hollow, called French drains are very general, even among the common farmers: some done with stones, but much with sods, laid on edge in the ground, they dig them 2½ or 3 feet deep'.[5] His note on drainage in the Barony of Forth, county Wexford, is probably more typical, however: 'They drain only with open cuts, no hollow ones done'.[6] By the beginning of the nineteenth century, agriculturalists were advocating scientific 'thorough' or covered drainage as essential for good husbandry. The methods developed by the Englishman, Joseph Elkington of Warwickshire, were generally agreed to be the best available. A key part of Elkington's technique was to release underground springs, using an auger or borer (Fig. 4). A typical application of the method was recorded in county Dublin, in 1801:

> George Grierson, Esq. has had some ground drained in the Parish of Tallaght, this year, according to Mr Elkington's mode, which appears to answer perfectly well; common drains were first opened to a depth of about two feet, and then bored at every eight or ten yards with a three inch auger; about six or seven feet deep; some of the holes, when I saw them, were throwing off a quantity of water, while others were dry, but I was informed, had run.[7]

Another account, published in 1802, claimed that Elkington's methods

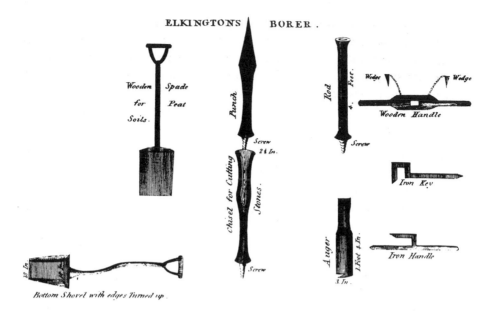

Fig. 4. A set of drainage implements, including Elkington's borer or auger for releasing underground springs (Coote, C. (1802), p.118).

had been 'tried in several parts of Ireland, and always with great success'.[8] Despite these projects, however, observers in most parts of Ireland lamented the lack of application of new principles. McEvoy's comments on county Tyrone in 1802 probably described common practice throughout much of the country: 'Open drains are in common use, only temporary to save crops in most situations, when the latter end of the spring is wet. The secret of hollow draining is very little understood in any part of the county, much less the intercepting, or cutting off springs. Sod drains are not known'.[9] Hely Dutton, writing just over twenty years later, repeated criticisms of what he saw as the haphazard drainage methods employed by most Irish farmers. He also warned, however, that where systematic drainage was attempted, proprietors should ensure that they engaged a *bona fide* expert:

> A few years since, Mr Elkington, nephew to the celebrated drainer in England came over to this country under an engagement to a few spirited gentlemen...; this gave such an impulse to draining that hopes were very generally entertained that great and permanent advantages would accrue to Ireland; certainly his works were excellent, and more neatly executed than the general practice had been, but [his methods]... were little or nothing different from those in practice by every other *scientific* drainer... Since

that period many *itinerant quacks* have started up in this branch... whose low terms have blinded the judgement of some landed proprietors, that in this, as in landscape gardening, have mistaken *neatness of execution for correctness of design*... Mr Hill, a native of north Britain, followed Mr Elkington; he was imported by the Farming Society of Ireland, and as far as I can judge... is an excellent drainer... since that period I have not heard of any person of eminence.[10]

By the 1840s, the drainage system advocated by the Scot, James Smith of Deanston,[11] was becoming widely known in Ireland. The Devon Commission found that on most large-scale schemes, Smith's plan of closed drains placed twenty-one feet apart, at a depth of two feet six inches, was being followed. Small drains, running parallel to one another, fed into larger 'sub-drains', which in turn discharged into main drains. Smith's system was not always followed rigorously, however. One deviation commented on by the Commission was the practice of running drains diagonally across slopes, rather than straight up and down them. The Commissioners were confident that this would soon be recognised as 'erroneous', as 'it admits of a simple mathematical proof'.[12] The diagonal layout was defended by some writers, however, who argued that on steep inclines, drains running straight down the slope would produce a current so strong that it would carry fine top soil away, and also possibly overdrain the ground.[13]

Smith's parallel lay-out of drains at a fixed depth was intended to be applied without alteration to any land, irrespective of small local changes in slope, soil type, or moisture content. The modified version of the system, devised by Josiah Parkes, which entailed fewer drains, laid deeper, also used a rigid parallel lay-out. It was this rigidity, rather than a rejection of the recommended techniques of drain construction, which led to a decline in the use of both systems. The point was clearly made in an Irish agricultural text published in 1863:

> The merits of the two systems formed a subject of dispute a few years ago, the great mistake being, that many on both sides considered it was possible that one universal rule was applicable in all cases; ultimately, however, people began to see that, owing to the diversified character of the soils, subsoils, and substrata of the land in this country, it was impossible to fix any one uniform depth or distance apart for the drainage of soils of every description.[14]

By the late nineteenth century, attempts to produce a completely standardised system had given way to the notion of 'occasional drainage'.

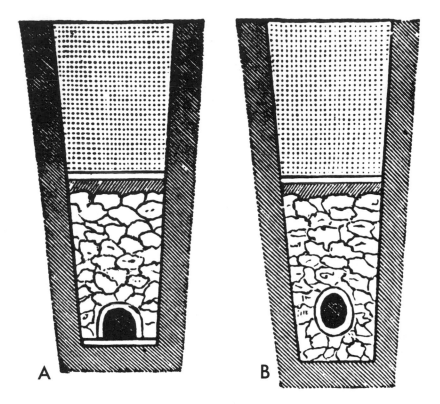

Fig. 5. A. 'Horse-shoe' drainage tiles set on flat 'sole' tiles which have been placed along the bottom of a drain. B. Cylindrical drainage tiles, commercially developed during the 1840s, and in widespread use by the 1870s (Doyle, M. (1844), p.204).

Experts were now expected to respond to small variations in soil, slope and waterlogging within the area to be drained.[15] This represented, in a more systematic form, something of a return to the approach associated with Joseph Elkington. It had been claimed that Elkington had become a famous drainer, not so much because of his scientific mastery of the principles of drainage, but because he was intuitively aware of the individual requirements of any area on which he worked.[16] Similar sensitivity began to be expected of more modern drainers.

One important change in drainage techniques, which occurred in the mid-nineteenth century, was the increasing use of clay tiles and pipes in the construction of drains. The earliest tiles to be widely adopted were known as 'horse shoes', which were placed on 'sole' tiles laid along the bottom of the drain (Fig 5a).[17] In 1843, however, at Derby Agricultural Fair, John Reade exhibited machinery for making cylindrical tiles.[18]

These tiles (Fig 5b) quickly came into widespread use, and by 1870 were regarded as the most common type in Ireland:

> Pipe tiles are the materials now most extensively used. They have been made of various sections, such as circular, horse-shoe shape, and oval. The section of pipe now most universally used is the circle. The diameter of the bore varies from one inch upwards. In the east of Ireland and where rainfall does not exceed thirty inches per annum, pipes an inch and a half in bore are quite capable of carrying off the water of parallel drains. In the south-west of Ireland, where the rainfall is double this, or where it falls in very heavy showers, two inch pipes should be used.[19]

By the end of the nineteenth century, the dominance of cylindrical tiles seems to have become complete. A British text, published in 1909, declared that 'horse-shoe' drains were 'a thing of the past'.[20]

After the Great Famine of the 1840s, large-scale drainage schemes were backed by government legislation and monetary grants. A Drainage Act had been passed in 1801 but proved ineffective. The Act of 1847, however, was much more successful. By 1850, over £3,500,000 in loans had been applied for under the terms of the Act, 20,000 labourers were employed in drainage, and 74,000 acres had been reclaimed.[21] The trend towards large-scale organisation increased during the later nineteenth century. The Famine and its aftermath had led to the removal, by starvation, emigration and eviction, of huge numbers of the poorest farmers, who had been the main agents of land improvement during the early part of the century.

Cultivation Ridges

At the start of this chapter, it was mentioned that cultivation ridges were not only central to reclamation techniques practised by the poorest farmers, but were also often integrated into large-scale schemes. The following general account of ridge-making will examine these claims in more detail.

Techniques of making ridges varied widely throughout Ireland, as also did their dimensions when complete (Fig. 6). The latter applied even when the same crop was being cultivated. In 1774, for example, Hyndman discussed the cultivation of flax on ridges ranging from three feet to twenty-one feet in width.[22] Very broad ridges were recorded in county Kilkenny in 1802, where some 'gentlemen' farmers practised a technique of ploughing 'in balk' for corn crops. The ridges made were

Fig. 6. Disused cultivation ridges on Clare island, Co. Mayo (photograph courtesy of Cambridge University Committee for Aerial Photography (ASY 16)).

up to sixteen yards wide.[23] In the mid-nineteenth century, however, the agriculturalist Martin Doyle claimed that the common Irish practice was to make 'extremely narrow' ridges, about four feet in width.[24] Unfortunately, the widths of furrows, or trenches, separating ridges were more rarely noted by observers. Arthur Young recorded examples of ridges six feet wide, which were separated in one instance by furrows two feet in width, and in another, two and half feet.[25] It seems to have been fairly common for narrower ridges to be accompanied by furrows which were between one third and one half of the ridges' breadth. Ridges also varied in height. The rather fragmentary evidence suggests that the centres of ridges might be anything from six inches to three feet higher than adjoining furrow bottoms.

Agriculturalists sometimes discussed the shape ridges should be in cross-section. Arthur Young believed that, in wet soils, ridges should be constructed so that in profile they would be smoothly rounded.[26] In 1802, McParlan described the construction of rounded ridges in county Leitrim on 'stiff tenacious soil . . . to give a fall of water into the furrows which they take care to keep clear'.[27] By contrast, however, some ridges made using spades were given firm, steep sides,[28] and in Erris, county Mayo, ridges were sometimes constructed to be asymmetrical in cross-section. One side of the ridge was built up as much as 50 cms. higher than the other. It was claimed that this helped drainage, and also that the sloping top of the ridge, which faced away from the prevailing wind direction, protected young plants.[29]

By 1800, the practice of making ridges in straight lines seems to have become widespread. Winding or 'curvilinear' ridges have typically been associated with medieval open-field systems.[30] However, in 1802, Tighe claimed that curved ridges were still frequently made in county Kilkenny,[31] and as late as 1844, Martin Doyle claimed that, 'The Common Irish, like the Norman farmer . . . often [makes] a winding ridge'.[32] Doyle believed that the curves in these ridges were simply the result of the ploughman following any bends in the sides of fields, and that there was no good reason for the practice. Elsewhere, however, he noted that, 'In many of the clay counties in England, the old-fashioned form of a gently winding ridge and furrow is still preserved'.[33] Here, he claimed, the intention was the same as that behind running drains obliquely across a slope 'to prevent a precipitate flow of the water'.

The practice of building up ridges on strips of untilled land was regarded as being particularly associated with Ireland. The widths of these strips varied. Very narrow strips might simply be the accidental by-product of ploughing. Ridges built up on strips which had been deliberately left unturned were known by various terms, including *iomairí*, rigs, or lazy-beds. Some of the techniques by which these ridges were made will be described in more detail in later chapters. The sensitive adjustments practised within these techniques in response to aspect, soil type, crops grown, their place in the crop rotation, and the methods of harvesting, show that, although developed by the poorest farmers, ridge-building cannot be simply thought of as an anachronistic and inefficient practice.

Ridges and Improvement

Ridge-making techniques were cited by many early agriculturalists as evidence that Irish agriculture was backward and inefficient. At Forth, in county Wexford, for example, Arthur Young found that one third of each ridge was left unturned during initial ploughing. He condemned this as 'execrable'.[34] Even Young, however, recorded that some 'improving' landlords were using lazy-bed techniques and finding them satisfactory. At Killarney, for example, he noted that, 'Mr Herbert has cultivated potatoes in the common lazy-bed method, upon an extensive scale, and he is convinced, from repeated experience, that there is no way in the world of managing the root that equals it'.[35] By the early nineteenth century, some observers were claiming that larger crops of potatoes could be produced from lazy-beds than from the newer system of drill cultivation.[36] The growing acceptance of lazy-bed techniques arose chiefly from the recognition of their usefulness in breaking in waste land for cultivation, where one potato crop could bring 'certain descriptions of ground into profit'.[37] Lazy-beds were particularly suited to the small-scale, labour-intensive methods available to cottars and small farmers, breaking in marginal land. They were also used in some large-scale reclamation schemes, however. In 1846, W.S. Trench planted 100 Irish acres of potatoes on mountain land in Queen's county (Offaly). The land was limed, and the outer edges of the lazy-beds turned by ploughing. The beds were then manured using guano, and covered with clay dug out of the furrows. Trench found that, 'The Potato grew to perfection in this rude description of tillage; and whilst it was growing, the heather rotted under the influence of the lime, and, together with the superabundant vegetable matter, was turned by the action of the lime into a most valuable manure'.[38] The advent of potato blight, however, brought a halt to this technique of reclaiming waste. This was lamented by the Devon Commission: 'The Irish or lazy-bed method of planting potatoes, supplied the most minute conceivable system of artificial draining for that one crop'.[39]

The acceptance of lazy-bed techniques as efficient was not so widespread in long-established tillage areas, however. By the beginning of the nineteenth century, some Irish farmers were experimenting with techniques of cultivation which did not involve the construction of ridges of any kind. One criticism of ridge cultivation was that the land taken up by furrows went to waste. As noted earlier, furrows could cover between one half and one third of the cultivated area.[40] It was also claimed that on rounded ridges the plants growing on the sides facing away from the light were 'shaded', and this hindered growth.[41] Tighe reported in 1802 that some farmers in county Kilkenny had attempted to cultivate wheat on level ground. This had not, however, been a success:

> This method was not found to answer as well as the common one: by means of the ridges in which wheat is usually placed, a greater surface of corn is exposed to the influence of the atmosphere, the ears are not so sheltered by each other from the air, nor so shaded from the light as they would be if the

field was even: that the air should circulate freely about the ears, is known to be an essential point with regard to wheat; a contrary situation encourages blights and mildew; . . . and though corn might not grow in the furrows between the ridges, yet the roots receive nourishment from them.[42]

These arguments for and against ridges, however, dealt with aspects which were to some extent secondary. The central aim in making ridges was to achieve drainage. The first major impulse to level ridges, therefore, came with the spread of the large-scale schemes of thorough drainage discussed earlier. Smith of Deanston believed that his drainage system rendered ridges obsolete:

> He [Smith] alluded to the practice, which he said had existed from time immemorial, of throwing the land into ridges and furrows, and showed that, by the soil being washed from the tops of the ridges into the furrows, the higher parts of the field produced comparatively little crop . . . In thorough-drained land no drop of water should run on the surface in any direction, but should penetrate into the ground where it fell . . . Wherever the land was drained, it was necessary that the high ridges should be done away with, and the land laid perfectly level.[43]

In Ireland, however, the application of Smith's methods was not always accompanied by an immediate levelling of cultivation ridges. Instead, ridge and furrow systems were integrated with thorough-drainage to increase the latter's efficiency, especially in wet, marginal land. On the Glenfin estate in county Donegal, for example, the laying down of drains was followed by the building up of ridges five feet wide, separated by furrows of two feet. If the land was flat, the ridges were made to run at right angles to the drains. If the drains ran down a slope, the ridges were made at an oblique angle.[44]

It was mentioned earlier that some cultivation ridges, especially those made by spades for planting potatoes, were deliberately given firm, steep sides. These sides, known as *bruach*, brews, or brows, were recognised by some observers as possibly hindering the efficient operation of thorough drains: 'The loosened soil at the bottom of every potato trench becomes necessarily the line of filtration, and the harder brows forming the edges impede the lateral transit of the water to the drains, so that, in fact, a good deal of the filtration in that case would resemble what takes place in an undrained field'.[45] The solution suggested was not that the drains should be flattened, however, but that, as in the Glenfin example just given, they should be made to run at right angles to the drains, and that the bottom of each furrow should be made to slope downwards, so assisting the flow of water.

One factor leading to the short-term retention of cultivation ridges on recently drained ground was their use in achieving subsoiling. The breaking up of impervious clays below the land surface was a central part of Smith of Deanston's system. Smith devised a subsoil plough to achieve this, but Irish contemporaries granted only a limited usefulness to the implement. James Kelly, agriculturalist to the Longford Farming Society, was emphatic: 'There is no tenant that I know of in this country that would think of getting a subsoil plough, nor do I think it would be wise of them to do so'.[46] In 1845, Lambert summarised the reasons for this lack of enthusiasm:

> The subsoil plough, as now in use, was invented by the justly celebrated Mr Smith, of Deanston . . . but, though admirably adapted to good flat plough-lands, such as the carses of Scotland, [the implement] is suitable here only in similar cases, and to proprietors and extensive farmers who keep strong horses and powerful teams. It is altogether out of the reach of our peasantry; and there are few large farmers who are so systematically extensive in tillage as to have the horse power necessary for the work.[47]

Agriculturalists recognised that subsoiling could be achieved by spades and spade ridges, which were more suitable to Irish conditions. An impervious substratum of clay, commonly known as 'lacklea' in Ireland, was exposed and broken up in the process of making ridges. Ridges, therefore, could have the same effect as a subsoil plough, and produce equally striking results. In county Sligo in 1823, for example, it was found that, 'Some of the low and moory grounds with . . . [lacklea] within three or four inches of the surface, are rendered fertile, and free from stagnant water by being cultivated for a potato crop, the subjacent stratum being carefully cut through by the loy or spade'.[48]

The practice of 'reversing' ridges from year to year was well established by the mid-eighteenth century. This meant that when ridges were made on a piece of land for a second year, the new ridges were built over the previous year's furrows. Agriculturalists pointed out that by this method, subsoiling could be effectively achieved over a whole field within two or three years.[49]

Despite all the advantages claimed for ridge cultivation, however, the building up of high, narrow ridges on a large scale did decline in the later nineteenth century. There were several important factors leading to this decline. The integration of ridge-making and thorough-drainage was most effective when marginal land was being broken in for cultivation. Once land was fully reclaimed, however, the need for ridges diminished. Other changes in agricultural practice also led to the eventual elimination of ridges. These included the spread of drill machinery in potato cultivation, and the adoption of scythes and reaping machines as grain harvesting implements. The implications these changes had for ridge cultivation will be discussed in later chapters. For the Devon Commission, reporting in 1847, the main disadvantage of ridge cultivation was that the intensive labour required would have been better employed in constructing thorough drainage systems, so creating much more permanent improvement of land:

> We may certainly estimate the extra labour on an acre of lazy-bed potatoes, as compared with drill culture, at £1 10s., and this extra labour on two acres would have been nearly sufficient, on the average, to have thorough-drained one acre; assuming, therefore, the quantity of ridge potatoes in Ireland to have been one million of acres, the extra labour wasted in this crop annually would have permanently drained nearly half a million of acres per annum, or the labour thus wasted during the last twenty years might probably have drained all the productive land requiring drainage in Ireland.[50]

During the second half of the nineteenth century, all of the factors which made ridge cultivation less attractive to farmers grew in importance. However, even in the 1980s some small holders cultivating tiny potato patches still rely on ridges to produce excellent crops.

Manure and Fertilisers

Processes aimed at enriching soils could take place either before or after tillage. One practice frequently discussed in eighteenth and nineteenth-century agricultural texts was paring and burning. There were many variations in the techniques used in this process, but the basic operations were 'stripping off the surface sod, allowing it to dry, burning it, and then spreading the ash on the soil as a fertiliser'.[51] In the early eighteenth century the practice was widely condemned by landlords, and in 1743 the Irish parliament passed an act forbidding it.[52] This Act does not appear to have been very effective, however, and by the early nineteenth century condemnations had become a lot less sweeping. The change in attitude came with the recognition that, as with lazy-bed techniques, controlled paring and burning could considerably speed up land reclamation. Arthur Young, though often critical of the practice, made use of it in the reclamation scheme he organised on the southern side of the Galtee mountains.[53] Burning some types of heath, along with the top sods containing their roots, could reduce previously tough, poor vegetation to productive ash. By 1802, Charles Coote, like many observers, concluded that, 'The effects of burning land were not well understood when the legislature imposed the heavy penalty against this process'.[54] It was argued by some agriculturalists that it was not paring and burning which ruined the land, but the overcropping which often followed it. In 1807, Rawson, describing the practice in county Kildare, made this point clearly:

> The common practice of burning the whole surface [of upland] and then applying the entire ashes on the remnant of the soil, taking three or four exhausting crops, cannot be too much reprobated; by it the land is completely exhausted, and men say how injurious paring and burning is, not considering, that the injury lies in making an improvident use of ashes.[55]

In 1810, Townsend argued that the unattractive appearance of the land after paring and burning was often the source of mistaken claims that the process was injurious:

> Were [paring and burning] . . . attended by the evil consequences so frequently deplored, the lands of Kinalea [Co. Cork] would by this time be reduced to a state of infertility. The contrary, however, is the fact . . . It is very probable that the naked appearance of land, let out without grass-seeds after burning, has been a principal cause of objection to the mode. But this barrenness is more apparent than real.[56]

The county statistical surveys published in the early years of the nineteenth century contain several descriptions of the use of paring and burning by landlords, some of whom had ploughs specially modifed for the purpose.[57] Modified forms of the process were also used in some of the most systematic schemes of land reclamation. In 1835, for example, one model project at Tullychar, county Tyrone, included the paring of sods, using the local type of breast-ploughs, known as flachters (Fig. 7).[58] Paring and burning continued throughout the later nineteenth century, but appears to have become increasingly confined to small farms in western counties such as Donegal,[59] and Mayo. In the latter case, the 'scraws' or top sods of an inch or an inch and a half thick were pared off, turned upside down until partially dry, and then 'footed' or built into small piles until they were completely dry, when they were then burned. The land from which the sods had been removed was made into ridges, and the ashes of the sods were spread over these. Potatoes were planted on top of the ashes, and covered with mould from the

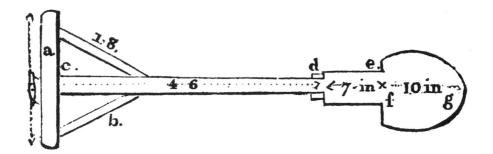

Fig. 7. A 'skrogoghe' or flachter from Co. Tyrone, used for paring off top sods (McEvoy, J. (1802), p.51).

furrows. The ashes were said to be 'Powerful for the plants', but it was also agreed that burning had 'spoiled a lot of land'.[60]

Unburnt clay and earth, dug elsewhere, were also commonly spread on cultivated land to increase fertility. The soil was sometimes spread straight on to the ground, or might first be mixed with farmyard manure.[61] Arthur Young recorded frequent instances of earth being taken from ditches to form composts, by both landlords and small farmers. In the Barony of Forth, county Wexford, for example, he found that, 'They are exceedingly attentive in getting mould out of the ditches and banks to mix a little dung with it, and spread it on their land'.[62] The large-scale removal of land from mountains and bogs was more controversial. The practice was widely reported, however. In Queen's county (Offaly) in the 1840s, it was reported that, 'In all parts [of the county] in which it can be obtained, even at a distance of several miles, bog soil, more commonly called 'bog stuff' or manure, is extensively used by large and small farmers, rich and poor . . . I know not what would have become of the population, or how they would have subsisted without its aid'. In some areas it was alleged that 'bog stuff' was used so much that the land on which it was spread had actually become moory.[63] Contemporaries were more generally worried by the detrimental effects on the land from which the soil had been removed, however. The removal of soil from marginal land was sometimes treated as theft:

On the ascent from Clogheen . . . in the Knockmeiledown mountains, we [investigators for the Devon Commission] met several carriers who live by a traffic . . . which they designate 'stealing mountain'. The stock in trade of this class is a donkey and cart. They derive their means of living from stripping the surface sod, *where one can be found*, from the unenclosed mountain, and thus accumulate compost, which they store at Clogheen as manure for potatoes for themselves, or to sell to the farmers. This system has been carried on in open day, in spite of the proprietors, for forty years. It is stated that 100 of these marauders have been summoned and fined at one session.[64]

The organised collection and storage of farmyard manure, was widely seen as neglected by agriculturalists. Arthur Young was emphatic:

> In the catalogue of manures, I wish I could add the composts found in well littered farm yards, but there is not any part of husbandry in the kingdom more neglected that this; indeed I have scarce anywhere seen the least vestige of such a convenience as a yard surrounded with offices for the winter shelter, and feeding of cattle.[65]

As suggested by this quotation, the apparent lack of attention to manure heaps can be linked to the lack of planned farm yards and outbuildings for the winter housing of livestock, rather than to ignorance of the importance of dung as a fertiliser. Where animals were housed, animal manure was efficiently collected. This can be seen even in the extreme case of the 'byre dwellings' described by a school teacher in an account of life in Gweedore, county Donegal, in 1837. In these houses, people lived under the same roof as their cattle, the two halves of the room being separated only by a drain which let some liquid effluent run out of the house (Fig. 8). It was claimed that ten or even fifteen tons of animal dung might be removed from these houses, when they were cleared out annually.[66] By 1856, government loans had been made available to farmers wishing to build planned farm yards. These included provision for liquid manure tanks.[67] Even in the later nineteenth century, the neglect of manure heaps was still criticised, however. Baldwin wrote that,

> In small farm districts it too frequently happens that the manure is accumulated in front of the door of the dwelling house, giving rise to the offensive effluvia, which often produce disease. Again, we often see small heaps of manure lying loosely here and there near the house and offices, by which its substance is wasted.[68]

Baldwin suggested that manure should be heaped on the north side of the farm yard and covered with a layer of absorbent peat or earth. Peat, he claimed, was best because animal effluvia speeded up its decay, and made it too into good manure. It can be argued, however, that the commonly reported practice of placing the dung heap in front of the dwelling-house door showed the importance attached to animal manure, rather than its neglect, a claim strengthened by the custom, in some

Fig. 8. The interior of a single-roomed byre dwelling from Magheragallen in north-west Co. Donegal, re-erected in the grounds of the Ulster Folk and Transport Museum (UFTM L1236/3).

areas, of planting ceremonial May-bushes on top of the heap.[69] The systematic care of manure heaps developed as farm yards became more formally laid out. However, this did happen only very recently on small farms in some areas. As late as 1892, it was reported from west Donegal that, 'cattle in many instances are housed at night at one end of the dayroom [of a dwelling], and the poultry often perch overhead'.[70]

In the mid-nineteenth century, the variety of manures used on a large scale increased significantly. Peruvian guano was widely used, as were ground animal bones, and, in the later part of century, the Irish co-operative movement gained some of its most striking successes in the large-scale trade in artificial manures. Throughout the eighteenth and nineteenth centuries, seaweed was widely used as a manure, especially for potatoes. Arthur Young claimed that at Westport, county Mayo, the effects of seaweed were so strong that the small farmers would 'not be at the trouble to carry out their own dung hills . . . A load of wrack is worth, at least, six loads of dung'.[71] In the barony of Tyrawley, he found that, 'The quantity of tillage is very inconsiderable, but what there is, is

Fig. 9. Kelp beds laid out along the shore of Carlingford Lough (UFTM L39/2).

vastly improved by the use of seaweed. Lands near the sea let at 20s. which at two miles, would yield but 14s. merely from being too far, as they reckon, to carry the seaweed'.[72] There are many accounts in nineteenth-century writings of the excitement with which people rushed to the shore to collect weed washed ashore during storms:

> Next morning the tempest was still high, and venturing upon the strand, I saw there, as a Valentia, crowds of females busied; and speaking to one, she replied, 'These stawrmy nights, ma'am, blow good luck to the poor; they wash up the say-weed, and that's why you see so many now at work'.
> The company increased, till I counted more than sixty; and busy, merry work they made of it; running with heavy loads upon their heads, dripping with wet, exultingly throwing them down, and bounding away in glee . . . 'And are you not cold?' 'Oh no, ma'am the salt say keeps us warm . . .' 'And how many days must you work in this way before you get a supply?' 'Aw, sometimes not fawrty, but scores of days'. 'And all you have for your labour is the potato?' 'That's all, ma'am, that's all . . .'[73]

Much more systematic use of seaweed was also common, however.

Fig. 10. Cutting seaweed on Inishere island, Co. Galway (photograph T. Mason, reprod. by courtesy of the National Museum of Ireland).

Farms near coasts often had rights to the weed growing on particular strips of beach. In 1837, Binns claimed that in county Kerry, rights were limited to the shore and rocks, but that weed growing under water was free to any one.[74] In other areas, however, 'kelp beds' ran far out into the sea. These were clearly marked off by lines of stones, which are still visible on some aerial photographs (Fig. 9). Within these beds, stones were sometimes laid down in sandy areas to encourage the growth of seaweed.[75] (This appears also to have been practised in more ancient times. The practice is implied in the Brehon laws, where it was intended to encourage the growth of dulse.)[76] In some cases rights to seaweed were sufficiently formalised to be stated in writing.[77] Rights to areas of seaweed persisted into the present century. On the Hornhead estate in county Donegal, for example, it was said that tenants knew the limits of the patches belonging to particular farms. Boundaries were not rigidly adhered to when the weed was plentiful, but disputes could arise if a farmer trespassed on to another's patch if the plants were not abundant.[78] Techniques of cutting growing weed varied. A long pole with a blade at one end was often used, either from a boat, or from rocks at the sea's edge (Fig. 10). The gathered weed was sometimes washed, a practice criticised by Lord George Hill,[79] and it was usual to dry the weed before carrying it to the fields. Much of the seaweed was burnt as kelp. This could be sold for chemicals, but a large proportion of the ash was also used as fertiliser.

Agriculturalists were generally agreed that seaweed, either dried or burnt, was an extremely effective manure. Its effects rarely lasted more than one season, however. The benefits from liming, on the other hand, could last for up to eight or nine years. Apart from limestone rock itself, other materials containing lime were also exploited. Marl clays were an important source. In 1863, marl was described as 'a calcareous earth, of which different specimens show very different amounts of lime in their composition . . . It is variously designated as clay, stone, and shell marl'.[80] Arthur Young recorded the dredging up of 'shell-marle' in Waterford harbour, and from the Shannon.[81] In 1808, Dutton recorded that,

> Astonishing improvements have been made in the neighbourhood of Killaloe, especially in the mountains between that and Broadford, by means of marle, inexhaustible quantities of which may be procured in the Shannon. It is raised by boats and drawn into heaps on the shore, where it generally lies until dry, and at leisure times is drawn to the land; about fifty loads are used to the acre.[82]

Marl clays were found in many parts of Ireland, especially under bogs. It was advised by Sproule, writing in 1839, that since the clay did not act so quickly on the ground as lime, it should be applied in larger quantities. He particularly advised its use on 'gravelly, sandy and peaty' soils.[83] Despite the recognition of the value of marl, however, it does not seem to have been exploited as extensively as other sources of lime. In coastal areas, shells and shelly sand were widely used, either being burnt or spread directly on to the land. Shells were identified as an important resource around lough Foyle as early as 1708,[84] and their use in the north-west has continued until very recently. In 1938, for example, shells were still being burnt for fertiliser on Tory island, county Donegal.[85]

One source of lime, claimed by several writers to be almost uniquely Irish, was limestone gravel. Arthur Young described this as 'a blue gravel, mixed with stones as large as a man's fist, and sometimes with a clay loam; but the whole mass has a very strong effervescence with acid. On uncultivated lands it has the same wonderful effects as lime, and on clay lands, a much greater'.[86] In 1802, Tighe claimed that, 'Before the practice of burning lime came into common use [in county Kilkenny], which is not above seventy years . . . [limestone gravels and sands] formed the principal manures of the county: for which reason great pits and excavations are to be found from whence they were raised'.[87] In 1824, Dutton also testified to the widespread use of gravel in county Galway: 'Limestone gravel has been formerly used in this county to such extent, that there is scarcely a field that has not an old gravel pit, and in some places there are several'.[88]

Arthur Young identified the large-scale burning of lime in kilns as a mid-eighteenth century development in several parts of Ireland.[89] He recorded a particularly rapid increase in the number of kilns in the area around Portaferry, county Down: 'There are many kilns . . . I was told 35, and that 15 years ago [circa. 1765] there was only one, so much is the improvement of land increasing'.[90] Lord Chief Baron Forster, Young's 'Great Improver', believed that his success in improving land was principally due to the use of lime:

> He had for several years 27 lime-kilns burning stone, which was brought four miles, with culm from Milford Haven. He has 450 cars employed by these kilns . . . The stone was quarried by 60 to 80 men . . . He spread from 140 to 170 barrels of lime per acre, proportioning the quantity to the mould or clay which the plough turned up. For experiment he tried as far as 300 barrels, and always found that the greater the quantity, the greater the improvement.[91]

The construction of lime kilns was widely discussed by early nineteenth-century agriculturalists (Fig. 11). In 1802, Tighe recommended that if kilns were not entirely made from limestone, at least the inner surfaces should be given a lining of the rock. This, if necessary, could be replaced every twelve years. The limestone lining would 'avoid vitrification', although this could also be achieved by using clay or unburnt brick.[92] Doyle, writing in 1842, recommended that kilns should be built of stone, lined with 'fire-brick'. He also argued that kilns should be narrow, as this led to economy of fuel. In tall, narrow kilns Doyle claimed, 'more than two thirds of the contents . . . [can be] drawn . . . every day, whereas . . . large circular kilns yield but half their contents in the day, and require considerably more fuel'. However, Doyle also pointed out that many kilns were constructed very crudely, and in many areas, especially in the south and west of Ireland, lime was burnt without using a kiln at all: 'A rude bank thrown up in a temporary way, on the side of any abrupt elevation of ground, answers the purpose of a regularly constructed building'.[93]

The local burning of lime continued on a large scale into the present century. One recent account from county Tyrone gives a clear description of the process:

IMPROVED LIME KILN,

INVENTED BY

THOMAS JAMES RAWSON, ESQ.

OF CARDINGTON,

IN THE

COUNTY OF KILDARE.

To face page 68—County of Cavan Survey.

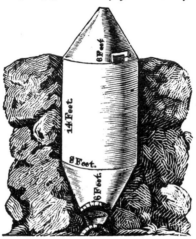

A lime-kiln fhould be made as high as the fituation of the ground will admit ; 20 feet is better than 16, 30 better than 20. The fides fhould be perpendicular. The annexed view is for 20 feet high ; the proportions fhould vary with the height. At bottom a metal plate with holes, fhould be placed fix inches above the lower part, to admit air, and for the fhovel to run on in drawing.

The drawing part fhould be fix feet ; width, 8 feet ; perpendicular fides, 14 feet ; on the head a cap is placed, formed like an extinguifher, brought to a hole at top of 12 inches diameter ; in the fide of the cap an iron door, with a latch is placed, to admit the charging the kiln, and to be kept clofe fhut. A kiln built on this plan will burn 9 barrels of lime for each of culm, and any fized ftones may be thrown in. Two active labourers muft attend it.

Fig. 11. A lime kiln invented by T. J. Rawson of Cardington, Co. Kildare (Coote, C. (1802), p.68).

One cartload would be a sufficient charge for a kiln. The stone was broken to about three-quarters of an inch mesh. The kiln was charged with alternate layers of peat and stone. The fire was lighted one evening, and during the next two days and one night some one must be constantly in attendance to stoke, first with approximately equal proportions of peat and limestone, and later with peat alone. The peat was added down the chimney. After forty-eight hours the fire was allowed to die, and the lime was raked out through the door.[94]

In 1839, Sproule listed some of the different ways in which lime was applied to land in Ireland:

1. It may be laid on the surface of land which is in grass, and remain there until the land is ploughed up for tillage . . .
2. It may be spread upon the surface while plants are growing . . .
3. It may be, and most frequently is, applied during the season in which the land is in fallow, or in preparation for what are termed fallow crops . . .
4. It may be mixed with earthy matters, particularly with those containing vegetable remains, and in this case it forms a compost.[95]

All of these methods of applying lime had been recorded by Arthur Young in the late eighteenth century,[96] and continued into the nineteenth century. The decline of 'bare fallowing', however, meant that liming associated with the practice also diminished in importance. Also, although Sproule specifically advised that lime should be mixed only with earth and not dung, mixtures of all three materials were very widely recorded.

Arthur Young, who was usually very critical of Irish cultivation techniques, was unreservedly enthusiastic about the amount of lime being produced from kilns: 'To do the gentlemen of that country justice, they understand this branch of husbandry very well, and practice it with uncommon spirit. Their kilns are the best I have anywhere seen, and great numbers are kept going the whole year through'.[97] Young was aware of the possibility of land being over-limed, but claimed that he had not found this anywhere in Ireland.[98] By the beginning of the nineteenth century, however, some agriculturalists were identifying over-liming as a growing problem. In county Armagh, Coote found lime to be 'prejudicial on light soils'.[99] He alleged that tenant farmers practised liming, 'when they mean to work the soil to the utmost it can produce, or, in other words, to run it out, previously to the expiration of a lease which they do not expect will be renewed'.[100] The changing attitudes to liming in the late eighteenth and early nineteenth centuries can be directly contrasted with those to paring and burning discussed earlier. Paring and burning was at first strongly condemned, but later gained a limited acceptance among agriculturalists as a technique for effectively breaking in some kinds of marginal land. Liming, on the other hand, was at first enthusiastically accepted, but was later treated with more caution. Lavish liming, like paring and burning, was defended as a practice where marginal land was being brought into cultivation. However, Dubordieu's summary of the effects of the over-use of marl and lime echoes contemporary descriptions of the effects of paring and

burning: 'From the imprudent use of this manure [marl], the land
became exhausted in some years, and fell off in its produce very
much . . . Marle has the same effect as lime . . . of enriching land, and,
like it, the effect of calling out its productive qualities to its
destruction'.[101] Similar effects were identified by Greig on the Gosford
estates in county Armagh, where upland pastures which had been limed
and tilled during the Napoleonic wars were eventually exhausted by the
lime's having brought 'every particle of fertility' of the land into
action.[102] It has been suggested that overliming was widespread in the
early nineteenth century.[103] In 1863, *Purdon's practical farmer* stressed
how long-term the injurious effects of the practice could be:

> We have sometimes met with cases where it was almost a hopeless task to
> undertake the improvement of over-limed land, from the state of thorough
> exhaustion to which it had been reduced; so much so, indeed, that even
> after being pasured for between thirty and forty years the effects of the
> overdose were quite visible when the land was broken up. In a case of this
> kind, the soil when ploughed feels quite puffy and dead under foot; and
> although oats will braird freely, and seem at first to be very promising, yet
> in a short time the leaves become brown, and the plants rapidly die away.[104]

Conclusion

All the processes described in this chapter show that during the late
eighteenth and nineteenth centuries the systematic application of new
ideas was gaining ground among large landowners and strong farmers.
With this went an increasing awareness among agriculturalists that the
techniques associated with a particular mode of cultivation were
interlinked and that ignoring this could produce long-term declines in
fertility. The integration of cultivation ridges and large-scale drainage
schemes suggests that by the mid-nineteenth century some landowners
were aware that not all old practices were bad, and that not all new ideas
should be followed to extremes. These developments were made explicit
in the writings of contemporary Irish agriculturalists. At the other end of
the scale, however, the smallest farmers also practised systems of
cultivation which could be sensitively balanced between gaining a short-
term subsistence living, and maintaining the long-term fertility of the
land. When, under pressure, these systems were upset, as for example in
the over-use of paring or burning or liming, individual small farming
families could, and did, face catastrophy.

NOTES

1. [Devon Commission], *Digest of evidence taken before Her Majesty's Commissioners of Inquiry into the state of the law and practice in respect to the occupation of land in Ireland*, part 1 (Dublin, 1847), p.571; Andrews, J.H., 'Limits of agricultural settlement in pre-Famine Ireland', in *Ireland and France, 17th-20th centuries*, eds. L.M. Cullen and F. Furet (Paris: École des Hautes Études, 1980).

2. Young, Arthur, *A tour in Ireland*, vol. 1 (Dublin, 1780), pp.146-150.

3. McEvoy, John, *Statistical survey of the county of Tyrone* (Dublin, 1802), p.190.

4. Young, Arthur, *op. cit.*, vol. 1, pp.369-381.

5. *Ibid.*, vol. 1, p.138.

6. *Ibid.*, vol. 1, p.111.

7. Archer, Joseph, *Statistical survey of the county Dublin* (Dublin, 1801), p.73.

8. Coote, C., *Statistical survey of the county of Cavan* (Dublin, 1802), p.120.

9. McEvoy, John, *op. cit.*, p.115.

10. Dutton, Hely, *Statistical and agricultural survey of the county of Galway* (Dublin, 1824), pp.169-170.

11. Smith, James, *Remarks on thorough drainage and deep ploughing* (Stirling, 1843).

12. [Devon Commission], *op, cit.*, pp.81-82.

13. Dutton, Hely, *op. cit.*, p.173.

14. *Purdon's practical farmer* (Dublin, 1863), pp.166-167.

15. Irish National Schools, *Introduction to practical farming*, vol. 2 (Dublin, 1898), p.362.

16. Wright, R.P. (ed.), *The Standard cyclopaedia of modern agriculture and rural economy*, vol. 4 (London, 1909), p.191.

17. Purdon, *op. cit.*, p.175.

18. Wright, R.P., *op. cit.*, vol. 4., p.192.

19. Baldwin, Thomas, *Handy book of small farm management* (Dublin, 1870), p.188.

20. Wright, R.P., *op. cit.*, vol. 4., p.217.

21. Connell, K.H., 'The colonization of waste land in Ireland, 1780-1845; *Economic history review*, vol. 3:1 (London, 1951), p.70.

22. Hyndman, C., *A new method of raising flax* (Belfast, 1774), p.15.

23. Tighe, W., *Statistical observations relative to the county of Kilkenny* (Dublin, 1802), p.201.

24. Doyle, Martin, *A cyclopaedia of practical husbandry*, rev. ed. W. Rham (London, 1844), p.496.

25. Young, Arthur, *op. cit.*, vol. 1, pp.219a and 33.

26. *Ibid.*, vol. 2, 'General observations', p.225.

27. McParlan, James, *Statistical survey of the county of Leitrim* (Dublin, 1802), p.50.

28. Bell, J., 'A contribution to the study of cultivation ridges in Ireland', *Journal of the Royal Society of Antiquaries of Ireland*, vol. 114 (Dublin, 1984), p.81.

29. Ó Danachair, Caiomhín, 'The spade in Ireland', *Béaloideas*, vol. 31 (Dublin, 1963).

30. Parry, M.L., 'A typology of cultivation ridges in southern Scotland; *Tools and tillage*, vol. 3:1 (Copenhagen, 1976), p.5.

31. Tighe, W., *op. cit.*, p.295.

32. Doyle, Martin, *op. cit.*, p.496.

33. *Ibid.*, p.195.

34. Young, Arthur, *op. cit.*, vol. 1, p.112.

35. *Ibid.*, vol. 2, pt. 1., p.117.

36. *Ulster farmer and mechanic* (Belfast, 1824-25), p.309.

37. Dubordieu, John, *Statistical survey of the county of Antrim* (Dublin, 1812), p.202.

38. Trench, W.S., *Realities of Irish life* (London, 1868), p.98.

39. [Devon Commission], *op. cit.*, p.79.

40. Hale, Thomas, *A compleat body of husbandry*, vol. 2 (Dublin, 1757), p.91.

41. Parry, M.L., *op. cit.*, p.12.

42. Tighe, W., *op. cit.*, p.184.

43. *The Farmer's Friend* (London, 1847), p.77.

44. [Devon Commission], *op. cit.*, p.100.

45. *Ibid.*, p.92.

46. [Devon Commission], *op. cit.*, p.116.

47. Lambert, Joseph, *Agricultural suggestions to the proprietors and peasantry of Ireland* (Dublin: Farmer's gazette, 1845), p.28.

48. North West of Ireland Society, *Magazine*, vol. 1 (Derry, 1823), p.87.

49. Lambert, Joseph, *op. cit.*, p.28.

50. [Devon Commission], *op. cit.*, pp.84-85.

51. Lucas, A.T., 'Paring and burning in Ireland: a preliminary survey;' in *The spade in northern and Atlantic Europe*, eds. R.A. Gailey and A. Fenton (Belfast: Ulster Folk Museum, 1970), p.99.

52. *Ibid.*, p.101.

53. Young, Arthur, *op. cit.*, vol. 2, 'General observations', p.69.

54. Coote, C., *op. cit.*, p.114.

55. Rawson, T.J., *Statistical survey of the county of Kildare* (Dublin, 1807), p.142.

56. Townsend, H., *Statistical survey of the county of Cork* (Dublin, 1810), pp.543-544.

57. Thompson, R., *Statistical survey of the county of Meath* (Dublin, 1802), pp.275-276; Townsend, H., *op. cit.*, p.124; Dubordieu, John, *Statistical survey of the county of Down* (Dublin, 1802), pp.175-176.

58. [Devon Commission], *op. cit.*, pp.610-611; Watson, M., 'Flachters: their construction and use on an Ulster peat bog', *Ulster folklife*, vol. 25 (Holywood, 1979).

59. Evans, E.E., *Irish folkways* (London, 1957 (1967)), p.147.

60. Lucas, A.T., *op. cit.*, p.112.

61. *Ordnance Survey Memoirs*, Box 26 II and VI, 1835.

62. Young, Arthur, *op. cit.*, vol. 1, p.112.

63. [Devon Commission], *op. cit.*, pp.66-67.

64. *Ibid.*, p.690.

65. Young, Arthur, *op. cit.*, vol. 2, 'General observations', p.68.

66. Hill, George, *Facts from Gweedore*, introd. E.E. Evans (Belfast: Queen's University, Institute of Irish Studies, 1971), p.17.

67. *Instructions to owners applying for loans for farm buildings* (Dublin, 1856), p.19.

68. Baldwin, Thomas, *Introduction to practical farming* (Dublin, 1877), p.6.

69. Evans, E.E., *op. cit.*, p.274.

70. Micks, W.L., *An account of the . . . Congested Districts Board for Ireland, from 1891 to 1923* (Dublin, 1925), p.253.

71. Young, Arthur, *op. cit.*, vol. 1, p.362.

72. *Ibid.*, vol. 1, p.347.

73. Nicholson, Asanath, *The Bible in Ireland.* Quoted in Frank O'Connor (ed.), *A book of Ireland* (Glasgow, 1959 (1974)), p.136.

74. Binns, J., *Miseries and beauties of Ireland*, vol. 2 (London, 1837), pp.131-132.

75. Young, Arthur, *op. cit.*, vol. 1, p.194.

76. Joyce, P.W., *A social history of ancient Ireland*, vol. 2 (Dublin, 1920), p.153.

77. Regulations for gathering seaweed, Ardglass, Co. Down. Public Records Office of Northern Ireland. Document T1009/160.

78. Testimony of Mr. Patrick Brogan, Hornhead (Ulster Folk and Transport Museum Tape, R79.36).

79. Hill, George, *op. cit.*, p.21.

80. Purdon, *op. cit.*, p.110.

81. Young, Arthur, *op. cit.*, vol. 2, 'General observations', p.68.

82. Dutton, Hely, *Statistical survey of the county of Clare* (Dublin, 1808), pp.157-158.

83. Sproule, John, *A treatise on agriculture* (Dublin, 1839), p.42.

84. Young, R.M., *Historical notes of old Belfast* (Belfast, 1896), p.159.

85. Mason, Thomas H., *The islands of Ireland*, 2nd ed. (London, 1938), p.6.

86. Young, Arthur, *op. cit.*, vol. 2, 'General observations', p.68.

87. Tighe, W., *op. cit.*, p.443.

88. Dutton, Hely, *A statistical and agricultural survey of the county of Galway* (Dublin, 1824), p.180.

89. Young, Arthur, *op. cit.*, vol. 1, pp.140 and 170.

90. *Ibid.*, vol. 1, p.189.

91. *Ibid.*, vol. 1, pp.147-148.

92. Tighe, W., *op. cit.*, p.440.

93. Doyle, Martin, *op. cit.*, pp.347-335.

94. Davies, O., 'Kilns for flax-drying and lime-burning', *Ulster Journal of Archeology*, 3rd Series, vol. 1 (Belfast, 1938), p.80.

95. Sproule, John, *op. cit.*, p.37.

96. Young, Arthur, *op. cit.*, vol. 1, pp.156, 9, 16, 34.

97. *Ibid.*, vol. 2, 'General observations', p.67.

98. *Ibid.*, p.68.

99. Coote, C., *Statistical survey of the county of Armagh* (Dublin, 1804), p.240.

100. *Ibid.*, p.239.

101. Dubordieu, John, *op. cit.*, 1802, p.183.

102. Greig, W., *General report on the Gosford estates in county Armagh, 1821*, introd. F.M.L. Thomspon and D. Tierney (Belfast: H.M.S.O., 1976), p.110.

103. Solar, Peter, 'Agricultural productivity and economic development in Ireland and Scotland in the early nineteenth century' in *Ireland and Scotland, 1600-1850*, eds. T.M. Devine and David Dickson (Edinburgh, 1983), p.79.

104. Purdon, *op. cit.*, pp.114-115.

CHAPTER 3

Spades

If we had to select the most distinctive Irish agricultural implement, the spade would be by far the most obvious choice. No clear evidence is available to show how long spades have been used in Ireland, but excavations of Bronze Age cultivation ridges suggest that spades or ploughs, or both, were used in ancient times.[1] Some iron-shod wooden spades which have been found in bogs and rivers may date from early Christian times (Fig. 12), although there has been some caution about ascribing a very early date to these. Both oral and documentary evidence suggests that similar spades may have been used as late as the nineteenth century.[2] During the last two centuries there have been more distinctively local types of spade than of any other Irish agricultural implement.

In the southern half of Ireland, most locally produced spades were made by blacksmiths. In the north, however, documentary evidence shows that by the 1760s specialised spade and shovel making mills were being established.[3] The northern concentration of spade mills was very marked. Sixty-seven spade mill sites have been identified throughout Ireland, and sixty of these were in the north.[4] Dublin and Cork were the major centres of large-scale production in Leinster and Munster. The numbers of spades produced by mills varied with the size of the mill, and through time. In 1830, one small mill at Aughrimderg in east county Tyrone was producing between 500 and 600 dozen spades a year. At the same period Carvill's mill in Newry was manufacturing 3,500 dozen spades and 57,000 shovels a year.[5] Although production from many mills seems to have been in decline by the 1880s, and by 1900 almost all of the very small specialised mills had gone out of production, there appears to have been an increasing use of spades manufactured in mills in areas where blacksmith-made implements had been most common. A retired blacksmith in county Cavan, for example, recalled having gone to MacMahon's mill near Clones during the First World War to show spade makers there the patterns of blades in south-west Ulster and north Connaught.[6]

Pattern books have survived from several spade mills, and these show

43

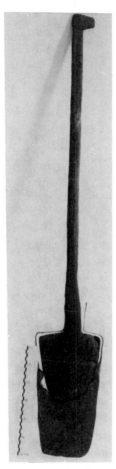

Fig. 12. An iron-shod wooden spade, found in Ulster (UFTM specimen 287:1967).

the great variety of spades produced. For example, one mill, from county Tyrone, had details of 230 different patterns. The blades ranged in weight from 3lbs. to 6lbs., in length from 13 to 22 inches. The width of blades at the mouth varied between 3 and 8 inches, and at the top, or shoulder, between 3½ and 8 inches.[7] Surviving specimens of blacksmith-made spades suggest that variations between these were also equally great.

Spade mill pattern books categorised spade types by area, and modern research has confirmed that there are clear regional differences in design. Some very distinctive local types have been recorded. For example, the 'gowl-gob' (Irish: *gabhal gob*) used in parts of county Mayo, had a two-pronged blade which was said to be useful in light sandy soils.[8] Several

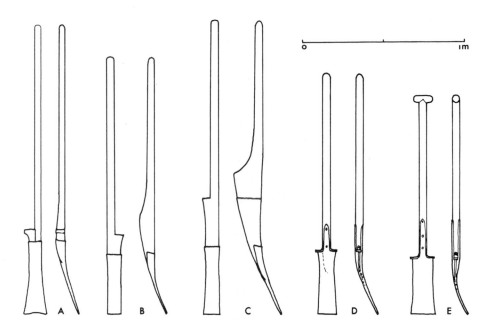

Fig. 13. A. A loy made by Scott of Cork, with a foot-rest made from a wedge of wood stuck into the spade socket alongside the shaft. B. A loy made by McMahon of Clones, with the shaft and foot-rest made from a single piece of wood. C. A 'big' loy made by Fisher of Newry. D. A two-sided spade by McMahon of Clones, with a long shaft, and the 'lift' formed by a crank in the blade. E. A two-sided spade made by McMahon of Clones, with a T-handle on the end of the shaft, and the 'lift' formed by a continuous curve in the blade.

broad distinctions can be made, however, which make the variety of spade types less overwhelming. The clearest of these is between one-sided and two-sided spades (Fig. 13). All Irish digging spades have foot rests at the top of their blades. Two-sided spades have two foot rests, one on each side of the shaft, while one-sided spades have only one. By the mid-nineteenth century, two-sided spades were found over most of north-eastern Ireland, while one-sided spades were used in the south and west.

One-sided spades are commonly known as loys (Irish: *láighe*), and have relatively narrow blades with open sockets. The shaft and wooden foot rest can be made from separate pieces of wood, the foot rest being a wedge stuck into the socket on one side of the shaft. This can be put either to the left or right, depending on which foot the spadesman uses for digging (Fig. 13a). This construction was found mostly in Munster and south Connaught. In south-west Ulster and north Connaught, on the

Fig. 14. Mr. John Dolan of Glengavlin, Co. Caven testing a 'big' loy for length. When the blade is set on the ground between the spadesman's feet, the upper end of the shaft should fit comfortably under his chin (UFTM, L296/19a).

other hand, the foot rest and shaft were often made from a single piece of wood (Fig. 13b). In parts of counties Cavan and Leitrim a massive version of this type of spade was used, known as the 'big loy'. The shafts of big loys were shaped to have a 'heel' of wood at the back of the 'loy-tree' which gave extra leverage when heavy sods were being turned (Fig. 13c). These implements were so heavy that it was usual for spadesmen to drag them along, rather than carry them.[9] Isaac Weld seems to have been discussing implements like these, when in 1832 he described loys in county Roscommon as 'hand-ploughs', which were no good for shovelling, but 'powerful and efficacious' in turning ground.[10] Like most one-sided spades, big loys had long shafts, which could be made either by local carpenters or the farmers themselves. Blacksmiths fitted the blade so that when the latter was set on the ground between the spadesman's feet, the upper end of the shaft fitted comfortably under his chin (Fig. 14).[11] Several nineteenth-century writers noted that the long shafts on many Irish spades meant that spadesmen could work in an almost perpendicular position. This also applied to some two-sided spades (Fig. 13d).[12] In north-east Ulster, however, many spades had shorter shafts which were fitted with T-shaped handles (Fig. 13e).[13]

In almost all Irish spades, the leverage was increased by an angle in the blade which made it curve towards the front. The angle or 'lift' might be a continuous curve (Fig. 13e), or might result from a sharp bend or 'crank' in the blade (Fig. 13d). In south-western Ireland some blades were bent so much that there was an angle of 35° to 40° between the lower part of the blade and the spade shaft.[14] As well as having a crank or lift, Irish spade blades were also dished, a feature useful in specialised spade work.

Techniques of Spade Work

Spade work was especially associated with techniques of building up ridges on strips of untilled or lea ground. As mentioned in the previous chapter, these ridges were commonly known as *iomairí* in Irish, but were referred to as 'lazy-beds' by most agriculturalists. Most of our detailed knowledge of lazy-bed techniques comes from material collected during fieldwork carried out this century. So far, this work has shown us how incomplete records are. It has also shown us just how refined spadework techniques have been, however. We can be fairly certain that the techniques recently recorded were also used in the nineteenth century. Although detailed documentary accounts of how spade ridges were made are very rare, passing references can sometimes be explained by modern practice. For example, a reference to 'sliding in the brews'[15] in one mid-nineteenth century account of spade cultivation can be understood, because the same technique is still used by some farmers in the west of county Fermanagh. (An account of how ridges are made in this area will be given below.)

Ridges were primarily made to assist drainage. We would expect, therefore, to find the most pronounced ridges in areas with badly drained soils and heavy rainfall. These conditions are common in the

Fig. 15. 'Making' land on the Aran islands (Mason, T. (1938), p.57).

west of Ireland, so it is not surprising that most fieldwork relating to ridges has been carried out in the west.

The technique of 'making land' on the Aran islands, county Galway, perhaps respresents the most extreme situation in which ridges have been made (Fig. 15). Thomas Mason described the basic processes in 1936:

> Alternative layers of sand and seaweed are placed on the bare rock, any crevices having previously been filled by broken stones to prevent the 'soil' from disappearing into them, for some of these fissures are ten feet deep; several layers, supplemented if possible by scrapings from the side of the road form the soil in which potatoes and other crops are grown. The ridges are made with a spade and one can hear this implement striking the underlying rock when an Aran man is working in his plot.[16]

Coarse sand was cspecially valued for making land, because when mixed with seaweed it could produce good crops. Grass also tended to grow best on coarse sand. Fine 'dead' sand tended to produce moss.[17] The labour of 'making' land was spent more in breaking rock and carrying

Fig. 16. Mr. John Dolan of Glengavlin, Co. Cavan making lazy-beds with a 'big' loy (UFTM, L296/15a).

sand and seaweed than in spadework. This can be contrasted with a method of making lazy-beds for potatoes using a 'big loy', which was recorded in Glengavlin, county Cavan, in the 1960s, and illustrates techniques used in many areas for making lazy-beds (Fig. 16).

Rows of farmyard manure were laid out on the grass to be tilled. In Glengavlin these rows were about 60 cms. wide, and 60 cms. apart. The spadesman used the edge of the loy blade to cut a line or 'score' down the middle of the grass strips between the rows of manure. The grass on each side of the score was then undercut to form sods which were turned over to lie on top of the manure on either side. The sods were undercut by forcing the loy blade into the turf at a fairly steep angle, and then pushing it under the sod at a much shallower angle. When the blade had cut right underneath the sod, the wooden 'heel' of the big loy was resting on the ground beneath. The loy was then levered to one side, so turning the sod over. The uncut side of the sod nearest the manure acted as a sort of hinge during this process. Sods turned to the right of the score were known as 'rip' sods, while those turned to the left were known as 'back' sods. Rip and back sods from either side of each strip of manure met to form a ridge, while the strips from which they had been turned formed the furrows. The potatoes were planted about one week after the ridges had been made.[18] In Glengavlin, an area of heavy wet soil, ridges and furrows were of narrow dimensions, whereas on drier soil, ridges were often made much wider. It was recorded in 1802, at Magilligan in county Derry, for example, that spade ridges were made up to eight feet broad 'on account of the dryness of the soil'.[19] Steeper slopes also sometimes meant that ridges were made wider, and in hilly areas around north-west Mayo, the upper, steeper parts of hillside fields were formed into wide ridges, which were then split into ridges only half as wide where the slope levelled off.[20].

Wider spade ridges were also made as lazy-beds, although in these the sods turned from the furrows did not meet in the middle of the ridge. The grass left in the middle of the ridge was covered with manure and loose soil dug from the furrows. Recent fieldwork, near Ederney in west Fermanagh, shows how complex techniques of making these lazy-beds could be (Fig. 17).

The spade used in making the lazy-beds (known locally as ridges) was of general 'Enniskillen' type.[21] As well as a spade a line made of hay rope, and a sally (willow) road, four and a half feet long (135 cms.), were used to mark the length and width of the ridge. Where no rod was at hand the length of the spade could be substituted as a measure for the width of the ridge. If the weather was dry, digging started in January, although February and early March was more usual. The line was laid along the field, near one side, to mark out the central line of the first furrow. The spadesman dug along one side of this line, taking out a narrow sod of earth, to a depth of about 5 cms., and the width of a spade blade. The line was then removed, and another row of sods of the same size was dug out alongside the first. The sods removed were thrown left and right to lie in the centre of the untilled area on which the ridges were to be built. These were broken up later, using the spade. Furrows were marked out in this way right across the field, at about 1.35 metres apart (Fig. 17a).

After several weeks sods were cut at the edge of the strips of grass between the furrows and turned to lie towards the centre of the ridge

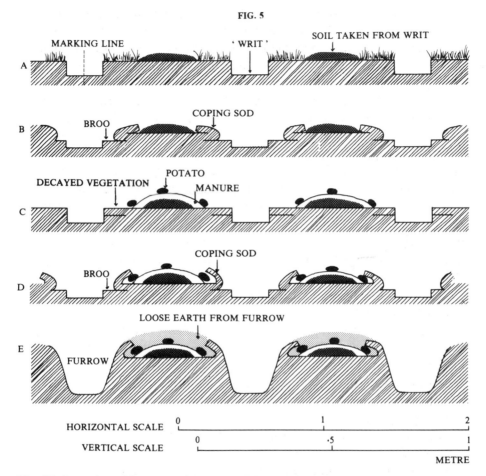

FIG. 5

Fig. 17. Stages in making spade ridges, near Ederney, Co. Fermanagh. A. Sods dug from furrows, and broken up in the centre of each ridge. B. Cope sods turned. C. Cope sods turned back again, manure spread, and potatoes planted. D. Cope sods turned again. E. 'Broos' sliced away, furrows deepened, and the loose soil piled on to the ridges.

(Fig. 17b). These 'cope' sods were thinner (about 2.5 cms.) than the sods dug from the furrows. The narrow shelf left between the furrow and the cope sods was known as the 'broo'. This word may be a dialect version of the English 'brow' or the Irish *bruach*. Both terms are used in documentary references to ridge cultivation,[22] as also is the term 'brew', mentioned earlier.

The 'cope' sods were flipped back into their original position just before potatoes were planted. By this time the grass on the sods, and the ground on which they had been lying, had begun to decay. Manure was then spread over the entire ridge, and potatoes were planted three abreast (Fig. 17c). After planting, the cope sods were again turned to lie over part of the ridge and partially covering the outer lines of potatoes (Fig. 17d). The broos on either side were then sliced away, and along with earth dug from the bottom of the furrows, this was piled on top of each ridge. The completed ridges were about 90 cms. broad and 20-35 cms. high. The furrows were about 45 cms. broad (Fig. 17e).

If potatoes were to be grown a second year, the old ridges were systematically dug away and new ridges built to run along the lines of the old furrows. Oats were also grown on ridges in west Fermanagh until very recent times, and if these followed potatoes, the old ridges were sliced away and the furrows filled in until the ground was almost level. Oat seed was then sown broadcast over the entire area. The seed was then covered with earth dug from along the central lines of the old ridges. This made new furrows which were circa. 22 cms. deep, but only 15 cms. wide, much narrower than when potatoes were being cultivated. Oats grown on ridges like these were harvested using reaping hooks or sickles. More recently, when oats were to be harvested using scythes, the old ridges were dug away until they were almost level, except for a 'slight waviness', the crests of the waves marking the centres of the old ridges. These were usually no more than 5 cms. higher than the level of the ground covering the old furrows. Some modifications were made in the dimensions of the ridges, depending on their situations. On low-lying, boggy ground, ridges were sometimes made as high as 30 cms., to prevent waterlogging. Since boggy ground also tended to dry out more during the summer, however, the sides of the ridges were made very firm. Strong sides were believed to hold moisture within the ridges.[23]

Techniques of making ridges varied widely, even within a small area. During fieldwork, Mr. Joe Kane of Ederney, who is expert in making ridges of the type just described, told us that when he and some neighbours went to help a man near Garrison, only twenty-five miles further west, to plant potatoes, they were amused and confused by both the type of spade used and the narrower ridges they had to make. In south-east Fermanagh, near Rosslea, ridges are made to almost the same dimensions as those recorded near Ederney, but here only a single sod is taken out to mark the furrows (Fig. 18a), and the 'cope' sods once turned (Fig. 18b) are left in position and the potatoes planted on top of them (Fig. 18c).

In some parts of Ireland, when ridges were being made in very wet soils, drainage was assisted by a thin layer of gorse or rushes spread along the ground on which the ridges were to be built.[24] Near Armoy in county

Fig. 18. Mr. Joe Flynn, from near Rosslea, Co. Fermanagh, making potato ridges in the grounds of the Ulster Folk and Transport Museum. A. A single row of sods is dug to mark out the furrow, and placed along the centre of an adjoining ridge. B. Thin cope sods are turned to mark the edges of each ridge. C. Manure is spread and the potatoes are planted. D. Soil is dug from the furrows to cover the potatoes (UFTM, L2442/5, L2443/6, L2444/1, L2444/4).

Antrim, the unturned strips of grass on which ridges were built were pierced with a digging fork at regular intervals to make them more pervious.[25] Further research into lazy-bed techniques will no doubt reveal many other ingenious refinements. The examples given here show that techniques were varied to allow for slope, soil-type, crops grown, and their place in the rotation.

The Efficiency of Spadework

During the nineteenth century in Ireland much debate occurred amongst agriculturalists as to the value of spade husbandry as a tillage technique. This included not only its application to small-scale farming but large-scale farming as well. Arguments ranged from the efficiency of particular Irish spade types as digging implements, and the best systems of tillage suited to Irish spades, to the more general question of the economic competitiveness of spade labour compared to horse ploughing. Many nineteenth-century agriculturalists were reluctant to admit that Irish spades could be efficient tillage implements. Their criticisms were especially directed against the one-sided loys. In 1808, Hely Dutton compared the loys used in county Clare with two-sided spades:

> The other implements are spades called in some parts of Ireland *loys* or *facks*. These are inconvenient heavy tools, throwing the weight on one hand, and greatly inferior to those in use in some parts of the county of Meath, and other parts of Leinster . . . they are . . . most unfit for cutting in bog, or for moving loose earth . . . the custom in this country, and indeed in most others, is to leave everything for the shovel: in arranging labourers, the stewards of this county allot a shovel to every spade, though a good shoveller could easily keep two spades employed.[26]

In 1834, Edmund Murphy argued that the narrow blades on most loys made then 'quite unfitted . . . for digging and trenching land'.[27] This criticism was repeated in 1870, when the upright position adopted by men using loys was also claimed to limit their efficiency.[28]

Some writers, however, recognised that the construction of loys made them well suited to working in heavy soils, especially when lazy-bed techniques were used. In 1802, McParlan praised the loys in use in counties Donegal and Leitrim:

> This sort of machine is admirably well adapted to the weight and tenacity of the soil. A broad short spade, pushed into this ponderous gluey stuff, must remain there, as if in a locked vice, whereas this narrow one fits to its own breadth a portable weight, and the long handle answering as a *lever* and the back . . . as a *fulcrum* in the operation of digging, very much facilitates the labour.[29]

One early nineteenth-century writer claimed that twelve men using loys could turn an acre in a day, which was twice the speed of digging with two-sided spades.[30] Although Thomas Baldwin criticised the upright position required when using the loy, he later argued that the loy for its specialised uses was 'an excellent implement cultivating the ground in a way superior to the most approved English spade'.[31]

The extent to which spade husbandry was seen as an efficient system of tillage and a viable alternative to horse ploughing was reflected in the holding of spade-digging contests to promote skilled spadesmanship. The match at the demesne farm of the Lord Primate of Ireland on 17th April, 1848 was held 'with a view to encourage spade husbandy'.[32] Contemporary reports suggest that these competitions were fiercely fought, and an account of the Rathdown Union Farming Society match in 1849 highlights the keen rivalry between spadesmen:

> The digging match took place in a field adjoining the ploughings. The ground allotted to each man, a statute square perch was dug in one hour and some minutes. There were 15 competitors, who went to work more like madmen than anything else. I need not say they had their coats off; but they had their hats off, and every part of clothing they could possibly do without. Two or three of them fainted as they finished their work; six of them had to be blooded since the competition; and others are reported to be dangerously ill . . . P.S. Since the above was written I have learned that one of the diggers at the match, an able, well conducted, good labourer named Grimes has died from its effects.[33]

Enthusiasm for the match at the Lord Primate's demesne farm was such that 'at the appointed hour, 105 well polished spades and digging grapes of different makes were on the ground'. Unfortunately, the size of the plot to be dug could only accommodate forty contestants.[34]

Practising farmers seem to have had few doubts about the value of spades as tillage implements. Arthur Young reported in 1780 that at Belleisle in county Fermanagh, crops of oats from fields prepared by spadework yielded so much more than those put in by the plough that even those farmers with horse plough teams preferred to plant their crops using spades.[35] In 1810, Townsend disapprovingly noted a similar preference in county Cork, when potatoes were being planted:

> The spade is the favourite implement of the farmer. This he principally employs in the culture of his most important, as well as most laborious crop, potatoes; though the soil, in general, is such as to admit the whole process to be, at least, as effectually, and much more cheaply performed by the plough. The necessitous circumstances of the cottager oblige him to resort to the only implement he can freely command, but the wealthier farmer, who has a plough and horses, half their time unemployed, is not so easily excused . . . An experienced agriculturalist cannot behold without surprise a farmer with half a dozen labourers, toiling for a fortnight (while his horses are doing nothing) to perform a piece of work, which his plough and harrow could accomplish in a couple of days.[36]

Townsend's own discussion, however, contains clues which allow an explanation of this apparently irrational behaviour. First, he admitted, the use of spades meant that the soil was well prepared for the crop. Second, despite a need for more manure, 'the bad method' was more productive!

In 1834, Edmund Murphy, whose condemnation of loys was mentioned earlier, argued that the spade 'of proper construction . . . is a much more efficient implement that the plough. The latter may turn over the ground, and the harrow may break the clods, but the spade turns the soil upside down . . . Every clod is broken, and weeds are deposited at a depth which at least checks their growth'.[37] By the later nineteenth century the 'deep, clean, and efficient' tillage possible with spades was claimed to have made a 5½ acre model farm at Glasnevin, county Dublin, 'one of the best tilled pieces of land in Ireland'.[38]

The main criticism voiced against spade husbandry for large-scale tillage was its profitability compared to horse ploughing. Critics of spade husbandry argued that, even if efficient as a means of breaking up the soil, the use of spades was both slow and labour-intensive and therefore more expensive. In 1834, the Scottish agriculturalist Low accepted that spadework was more efficient than ploughing, but argued that the greater expense meant that it was unlikely to be economic 'on the great scale, where the profit depends upon economy of labour'.[39] Supporters of spade husbandry, however, like 'J.J.G.' of Cashel, county Tipperary in 1847, claimed that while the initial costs of cultivating a large farm by spade labour were greater than with the plough, the subsequent gains resulting from better crops favoured spade husbandry. 'J.J.G.' calculated that the cost of tilling land by the spade was £6 per Irish acre or 1s. 6d. per day task work, whereas with two horses and a plough it was only £2.19s. 0d. However, he concluded that the difference of £3. 1s. would be 'made up and half as much more again', by the superior crops produced by spade culture.[40]

The problem of the economic competitiveness of spade husbandry, particularly for large-scale tillage, could also be diminished where there was abundant cheap labour. In the eighteenth century teams of spadesmen hired themselves to farmers. These 'spalpeens' (Irish: *spailpíní*) were especially associated with Munster.[41] In county Cork, there was an annual influx of temporary workers from western areas when potatoes were being dug and lifted.[42] In many parts of Ireland, however, there was no need to hire workers for spadework, either because holdings were very small, or help from neighbours ensured that fields could be planted and harvested quickly. A 'gathering' (Irish: *meitheal*) of neighbours could vary considerably in size. The most spectacular of these gatherings occurred in Ulster just before the 1798 rebellion. In one case it was claimed that 6,000 men had come together, insisting that they wanted to dig a poor woman's potato patch for her! This apparent manifestation of neighbourliness was, however, dismissed as a ruse by the authorities: 'The main object of the potato digging is probably to enable the leaders to ascertain how their men will act at the word of command'.[43]

Labour was so cheap in Ireland during the 1840s that one county Meath farmer concluded that since he paid a guinea to have an acre ploughed, whereas a spadesman was paid only 9d. or 10d. a day, 'I think I will dig it very nearly as cheap as plough it; and when done, it will be much better done'.[44] Spade labour used for large-scale tillage was regarded by some observers in the nineteenth century, not just as an economic and efficient method of cultivation, but as a means of reducing Ireland's wider social problem of rural unemployment and, consequently, the burden on the poor rate. E. Barry, a member of the Fermoy Union Farming Society, county Cork, complained in 1849 that he could not understand 'why our landed proprietors and farmers will not enter into spade husbandry', and calculatd that, 'by giving the work to those who be otherwise unemployed every acre dug at 1d. per perch will keep one hundred persons out of the workhouses'.[45]

A striking example of the use of the rural unemployed for large-scale spade husbandry was on the land of Colonel Knox Gore, near Ballina, county Mayo. Three hundred acres, neglected through emigration, were brought under cultivation in one season by 'the destitute of the Ballina Union'. It was reported that from December 1848 to July 1849, 'with untrained half-starved emaciated men there were 100 Irish acres dug 14 inches deep, sown with flax; 100 acres dug 10 inches deep, sown with oats; and 100 acres dug twice, drills opened, manure deposited, covered with the spade, and sown with turnips, which deducting all expenses of labour, manure and seed has left a profit'. The estimated profit on the venture was calculated to be £740.[46]

Spade husbandry was recommended by agriculturalists for deep cultivation, particularly in the early nineteenth century for the tillage technique known as trenching. Deepening the soil was thought to improve it, and when trenching was employed for the purpose, the land was dug in parallel trenches, bringing the subsoil to the surface.[47] Loudon described trenching as 'a mode of pulverising and mixing the soil or of pulverising and changing its surface'.[48] Pulverising and changing of the soil surface however, could mean that the relative positions of the strata of the soil were simply reversed, with the subsoil

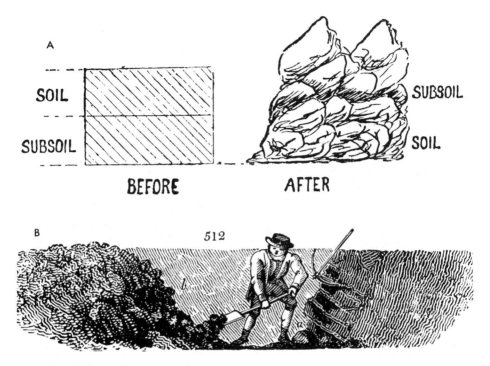

Fig. 19. A. Trenching, where the soil strata are inverted. The subsoil is put on top, and the surface soil beneath (Baldwin, T. (1898), p.369). B. Trenching, where the soil strata are mixed together. Note the depth to which the soil is dug (Loudon, J. C. (1831), p.508).

covering the soil (Fig. 19a). To mix and pulverise the soil effectively Loudon argued that the face or position of the moved soil in the trench should 'always be that of a slope in order that whatever is thrown there may be mixed, and not deposited in layers as in the other case'[49] (Fig. 19b).

Although the above methods of trenching advocated by theoretical agriculturalists were practised, the technique commonly referred to as trenching in Ireland, was not a preparatory tillage method, but was part of ridge cultivation used during the planting of the potato crop, and in the covering of sown cereal and flax seed. Wakefield roughly described trenching in Ireland as 'to form the land into beds and shovel out a deep trench between them, throwing up the earth in beds on each side'.[50]

The shovel (Fig. 20) was widely used in conjunction with the spade for trenching in ridge cultivation, and its function as an implement was to lift and spread broken up soil. Evans claims that most Irish shovels were

Fig. 20. Paddy and Cassie McCaughey of Ballymacan, Co. Tyrone making ridges, in 1906. Cassie is holding a long-handled shovel (UFTM, L918/4).

long-handled, and classifies them into two main types, round-pointed and square-pointed. It was said of the spade and shovel in county Armagh in 1804 that 'much work is done with the spade and the shovel, as they trench in most of their crops for this use; the shovel is square, and well adapted to the purpose'.[51] Shovels in more recent times have had all-metal blades, although in the mid-nineteenth century W.R. Wilde noted that 'until very recently the shovel common in Galway and other parts of the west of Ireland, was composed of wood, shod with iron round the edge for about 2 inches in breadth; it was usually made of sallow'.[52]

Trenching which involved the inversion of the soil strata was widely hailed by theorists in the late eighteenth and early nineteenth centuries as the correct method for all deep tillage. However, later nineteenth-century agriculturalists were less enthusiastic about the *carte blanche* application of the method, as it was found that the burying of some surface soils by the subsoil spoilt their fertility. An Irish National Schools agricultural text book of 1898 stressed that surface soil should not be buried in trenching and that care should be taken in bringing the subsoil to the surface. It was recommended that a subsoil which was a cold clay should not be brought to the surface. If the soil rested upon gravel or sand it did not require trenching, but if the soil was a strong clay or a peat and the subsoil of a sandy or gravelly character, a little of the subsoil could be mixed with soil.[53]

The changing attitude to trenching aptly illustrates the gulf between theoretical and practical agriculture in the nineteenth century. A particular system or agricultural principle was often propounded by theorists without regard for local conditions and often reflected gaps in the knowledge of the theorists themselves. Informed late nineteenth-century opinion pointed out that small organisms, which were found in the surface soil and aided the fertility of the soil, were killed if the surface soil was totally buried by trenching. This was obviously not fully understood by some earlier theoretical agriculturalists. It was noted in 1898 that in Ireland subsoiling had in most places taken the place of trenching.[54]

Whilst the suitability of spade husbandry for large-scale tillage was fiercely argued, it was widely accepted that it was best adapted to small-scale farming, as the following poem aptly illustrates:

> Let little farmers mind their spades,
> Nor think of keeping four-legged jades;
> The proverb long ago decides
> Which way a mounted beggar rides.[55]

Farms of under ten acres were generally thought to be of a size which could most efficiently use spade husbandry. Thomas Baldwin, discussing the 5½ acre model farm at Glasnevin, argued that despite worries amongst economists about the viability of such tiny holdings, the use of spade work, proper rotations, and stall feeding of animals, could provide full-time labour for a farmer and one of his sons, and keep the farm family in modest comfort.[56] By the early twentieth century, spadework seems increasingly to have become confined to the cultivation of vegetable patches, or to farms where the pastoral element was dominant. However, even in the 1940s, it was found that in county Fermanagh, 'On the smallest farms of 10 acres or less most of the work is done by spades . . . The Fermanagh or McMahon spade is well adapted to the thin soils of the county. It is a paring rather than a digging spade'.[57] Modern fieldworkers still find these western counties a very rich source of information on the complexities of spadework.

NOTES

1. Caulfield, Seamus, 'Neolithic fields: the Irish evidence' (B.A.R., 48), eds. H. Bowen and P. Fowler (Oxford, 1978).

2. Gailey, R.A., 'Irish iron-shod wooden spades', *Ulster journal of archaeology*, vol. 31 (Belfast, 1968), p.84; Ó Donachair, Caoímhín, 'The spade in Ireland', *Béaloideas*, vol. 31 (Dublin, 1963), p.112.

3. Gailey, R.A., *Spade making in Ireland* (Holywood: Ulster Folk and Transport Museum, 1982), pp.v-vi.

4. *Ibid.*, p.vi.

5. *Ibid.*, p.33

6. *Ibid.*, p.vi.

7. Evans, E.E., *Irish folkways* (London, 1957 (1967)), p.135.

8. Lucas, A.T., 'The "gowl-gob", an extinct spade type from county Mayo, Ireland', *Tools and tillage*, vol. 3:4 (Copenhagen, 1979).

9. Gailey R.A., 'Spade tillage in south-west Ulster and north Connaught; *Tools and tillage*, vol. 1:4 (Copenhagen, 1971).

10. Weld, Isaac, *Statistical survey of the county of Roscommon* (Dublin, 1832), pp.657-659.

11. Gailey, R.A., *op. cit.*, (1982), p.1.

12. Weld, Isaac, *op. cit.*, p.659; Binns, J., *Miseries and beauties of Ireland*, vol. 2 (London, 1837), p.215.

13. Gailey, R.A., 'The typology of the Irish spade', in *The spade in northern and Atlantic Europe*, eds. R.A. Gailey and A. Fenton (Belfast: Ulster Folk Museum, 1970), p.35.

14. *Ibid.*, p.39.

15. [Devon Commission], *Digest of evidence taken before Her Majesty's Commissioners of Inquiry into the state of law and practice in respect to the occupation of land in Ireland*, part 1 (Dublin, 1847), p.587.

16. Mason, Thomas H., *The islands of Ireland*, 2nd ed. (London, 1938), p.58.

17. Mullen, Pat, *Man of Aran* (London, 1934), pp.20-21.

18. Gailey, R.A., *op.cit.* (1971), pp.229-233.

19. Sampson, G.V., *A Memoir, explanatory of the chart and survey of the county of London-derry, Ireland* (London, 1814), p.300.

20. Ó Danachair, Caoimhín, 'The use of the spade in Ireland', in *The spade in northern and Atlantic Europe*, eds. R.A. Gailey and A. Fenton (Belfast: Ulster Folk Museum, 1970), p.52.

21. Gailey, R.A., *op.cit.* (1982), p.5.

22. Stephens, Henry, *Book of the farm*, vol. 1, 3rd ed. (Edinburgh, 1871), p.94; Ó Donaill, N., *Foclóir Gaeilge-Béarla* (Dublin: Oifig an tSoláthair, 1977), p.146.

23. Bell, J., 'A contribution to the study of cultivation ridges in Ireland', *Journal of the Royal Society of Antiquaries of Ireland*, vol. 114 (Dublin, 1984).

24. Ó Danachair, Caoimhín, *op. cit.* (1970), p.54.

25. Interview with Mr. Archie Mullen, Armoy, county Antrim (Ulster Folk and Transport Museum tape R82.73).

26. Dutton, Hely, *Statistical survey of the county of Clare* (Dublin, 1802), pp.65-66.

27. Murphy, Edmund, 'Spade husbandry', *The Irish farmer's and gardener's magazine*, vol. 1 (Dublin, 1834), p.680.

28. Baldwin, Thomas, *Handy book of small farm management* (Dublin, 1870), p.25.

29. McParlan, James, *Statistical Survey of the county of Leitrim* (Dublin, 1802), p.30.

30. Evans, E.E., *op.cit.*, p.134.

31. Baldwin, Thomas, *Introduction to practical farming*, 23rd ed. (Dublin, 1893), p.148.

32. *The farmer's gazette*, vol. 7 (Dublin, 1848), p.197.

33. *Ibid.*, vol. 8 (Dublin, 1849), p.93.

34. *Ibid.*, vol. 7, p.197.

35. Young, Arthur, *A tour in Ireland*, vol. 1 (Dublin, 1780), p.275.

36. Townsend, H., *Statistical survey of the county of Cork* (Dublin, 1810), pp.246-247.

37. Murphy, Edmund, *op.cit.*, p.680.

38. Baldwin, Thomas, *Introduction to Irish farming* (London, 1874), pp.116, 123.

39. Low, D., *Elements of practical agriculture* (Edinburgh, 1834), p.160

40. *The farmer's gazette*, vol. 6 (Dublin, 1847), p.375.

41. Ó Danachair, Caoimhín, *op. cit.* (1963), p.98.

42. Townsend, H., *op.cit.*, p.616.

43. Salaman, R.N. *The history and social influence of the potato* (Cambridge, 1949 (1970)), p.269.

44. [Devon Commission], *op.cit.*, p.950.

45. *The farmer's gazette*, vol. 8 (Dublin, 1849), p.582.

46. *Ibid.*, p.582.

47. Irish National Schools, *Introduction to practical farming*, part 2 (Dublin, 1898), pp.368-369.

48. Loudon, J.C., *An encyclopaedia of agriculture* (London, 1831), p.508.

49. *Ibid.*

50. Wakefield, E., *An account of Ireland, statistical and political,* vol. 2 (Dublin, 1812), p.431.

51. Coote, C., *A statistical account of the county of Armagh* (Dublin, 1804), pp.177-178.

52. Wilde, W.R., *Catalogue of the antiquities of the museum of the Royal Irish Academy* (Dublin, 1857), pp.206-207.

53. Irish National Schools, *op.cit.,* vol. 2, pp.369-370.

54. *Ibid.,* p.371.

55. Doyle, Martin, *Hints originally intended for the small farmers of the county of Wexford* (Dublin, 1830), p.30.

56. Baldwin, Thomas, *op. cit.* (1874), pp.123-124.

57. Mogey, J.M., *Rural life in northern Ireland* (London, 1947), p.41.

CHAPTER 4

Ploughs

Irish ploughs of the late eighteenth and early nineteenth centuries (Fig. 21) were probably more criticised than any other 'common' implement. A typical description, given by Martin Doyle in 1844, illustrates how unimpressed agriculturalists in general were:

> The share is like a large wedge; the coulter comes before the point of the share sometimes, and sometimes stands above it; the earth-board is a thing never thought of, but a stick (a hedge-stake or anything) is fastened from the right side of the heel of the share, and extends to the hind part of the plough; this is intended to turn the furrow, which it sometimes performs, sometimes not; so that a field ploughed with this machine looks as if a drove of swine had been moiling it [churning it up].[1]

If we examine the documentary evidence more closely, however, it is clear that there were in fact more types of 'common' Irish plough in use than was suggested by Doyle. Earlier in the nineteenth century another agricultural writer had said that there was such a variety of implements that no two counties in Ireland had ploughs of the same construction.[2]

The simplest plough to have survived in use during the last two centuries was of a type known internationally as an *ard*. A nineteenth-century landlord at St. Ernan's, county Donegal, described ploughs used on his estate which were constructed from 'a crooked stick with a second one grafted on to it, to give two handles, and the point of the stick armed with a piece of iron'. There is some evidence that similar ploughs were used in parts of county Fermanagh in the early decades of the present century.[3] However, despite the variety of ploughs, and the sketchy evidence we have to go on, it is possible to identify two main types among those which agriculturalists described as 'common'. These were the old Irish long-beamed plough, and the light Irish plough.

Old Irish Long-Beamed Ploughs

Ploughs may vary in minor constuctional details, but if the work they perform and the system of tillage in which they are used are the same,

Fig. 21. A wooden plough in use near Hillsborough, Co. Down, in 1783. The plough team of four horses, and the driver leading them, were common when heavy wooden ploughs were used. The collar on the nearest horse appears to be made of súgán, or plaited straw (detail from a print by W. Hincks).

the major working parts are likely to share similar characteristics. The old Irish long-beamed plough was a swing (i.e. wheelless) plough. Its other main characteristics were a long beam, flat wooden mouldboard, a long sole plate, the lack of a plate on the plough's land side, and its heavy weight. The long-beamed plough of county Kildare was claimed to be 'unwieldy . . . the weight sufficient to load two of the four animals, who are obliged to lug it'.[4] In 1802, the 'common' plough of county Kilkenny (Fig. 22) was described as 'clumsy' with a seven to nine feet long beam, a straight mouldboard, a sole with too long a heel, and a plough body without a side-plate.[5] In county Cork in 1812, 'common' ploughs were said to differ from the county Kilkenny ploughs only in minor details, and ploughs with similar characteristics were also reported in counties Armagh, Derry, Dublin, Meath and King's and Queen's.[6]

Comparisons between these ploughs are possible only where detailed descriptions have been provided. This is the case for the ploughs recorded in Meath, Kilkenny and Derry. A comparable plough has also recently been discovered in the roof of a pig-sty near Moira, county Down (Fig. 23). This has close similarities to the early nineteenth-

Common Plough.

Fig. 22. A long-beamed, heavy wooden plough from Co. Kilkenny (Tighe, W. (1802), p.294).

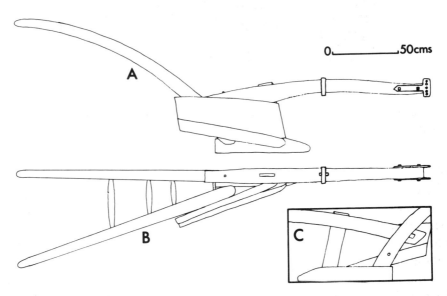

Fig. 23. A heavy wooden plough from near Moira, Co. Down. A. Side, showing mouldboard and reest. B. Plan. C. Detail of land-side (UFTM, L2318/1).

century types. The county Meath plough (Fig. 24) is distinct from the others for which we have information in having a mouldboard which was described by Thompson in 1802 as 'composed of three pieces, fastened to the cross [breast] and handle by wooden pins; the lowest is very small, and runs into the share, which is invariably made of cast iron; the next is somewhat broader and longer, and projects further in width; the third is considerably the broadest.'[7] The ploughs from Moira and county Kilkenny, by contrast, had mouldboards made from a single piece of wood. A distinctive feature of the Moira plough, however, is the long

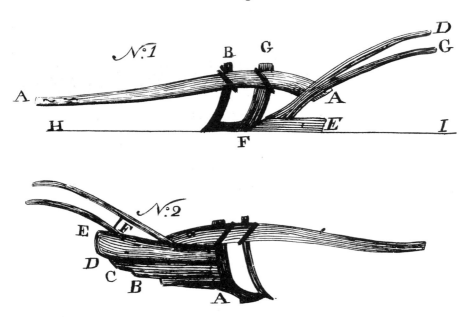

Fig. 24. A long-beamed, heavy wooden plough from Co. Meath (Thompson, R. (1802), p.108).

block of wood, known as a 'ground-wrest', which is fixed to the bottom of its mouldboard. It is possible that the Meath and Kilkenny ploughs had similar pieces of wood attached temporarily to their mouldboards, particularly for summer ploughing, but the ground-wrest on the Moira plough also links it, along with other features, to the 'Old Scotch' ploughs used at the same period.[8]

One important difference between the Moira plough and those from Derry, Meath and Kilkenny was that the latter three seem to have had adjustable beams. This construction was criticised as weakening the plough, but it did have a use. The system for linking draught harness to the 'common' plough from county Meath was described as 'two hooks on the beam point . . . to each of which a pair of horses is yoked'.[9] These fixed hooks did not allow any adjustment of the line of draught, which determined the depth at which the plough ran through the ground. Such an adjustment could be made, however, by raising or lowering the plough beam on the plough breast. On this plough, when adjusting the beam, the length of the plough's breast could be varied from between 18 and 22 inches.[10] In Derry, farmers adjusted the height of the beam by means of wedges in the mortice between the beam and the long handle.[11]

The adjustable beam served the same purpose, therefore, which was fulfilled on the Moira plough by the mechanism used to link the draught harness to the beam. The front end of this plough's beam was fitted with an iron 'muzzle' made from two T-shaped pieces of wrought iron, each punched with a vertical line of five holes. By connecting the harness to higher or lower holes in the muzzle, the line of draught was changed, so making the plough run shallower or deeper through the ground. The muzzle on the Moira plough was probably a product of eighteenth-century 'improved' design. Although it does not survive, the muzzle was probably fitted with an iron 'bridle', which had holes punched through it laterally. By attaching the harness to one of these holes, the ploughman could vary the strain of the draught from left to right, and so plough a wider or narrower furrow.[12] In England this mechanism was included on the famous Rotherham plough. As one writer pointed out, 'The difference betwixt this and the common ploughs, seems to consist in the bridle at the end of the beam, by which the ploughman can give the plough more or less land . . . or make her plow deeper or shallower.[13] On the Rotherham plough, the bridle and muzzle were a single fitting. Surviving nineteenth-century Irish ploughs, based on Scottish designs, have coupling systems on which the bridle is separate from the muzzle. At present the Moira plough seems to be one of the oldest surviving Irish ploughs on which the two-piece mechanism is found.

We can only decide on the efficiency of any plough by relating it to the sort of work for which it was used. This information is rather sparse for old Irish long-beamed ploughs, but in county Kilkenny at the beginning of the nineteenth century, the general work for which they were used was said to be 'breaking and turning fallows, ley ground or burnt ground'.[14] The old Irish long-beamed plough seems to have turned furrow-slices in a very particular way. In 1802, Richard Thompson noted that 'the share does not lie flat on the ground, but runs on its side, leaving the under surface, if all the loose clay was removed, cut into ridges somewhat resembling the letter V, or rather its transverse section, resembling the teeth of a saw'.[15] The long-beamed ploughs of counties Kilkenny and Dublin were said to have similar effects on the ground beneath the furrows. This distinctive pattern was created by the ploughman counter-acting a pull on the plough caused by the share being set with its point slightly toward the land-side. To compensate for this pull, the ploughman had to lean on the left handle of the plough, and opposite the cutting edge of the share.[16] At Castleknock, county Dublin, Hely Dutton claimed that 'when the plough was entered in the ground, the ploughman threw himself almost on his left side'.[17] By leaning on the plough in this way, the ploughman raised the cutting edge of the share, and so produced the ribbed furrow bottoms. The ribbed pattern was criticised by early agriculturalists because it left a tract of unploughed land at the bottom of each furrow slice. However, when a fashion for 'high crested' furrow slices developed in the latter half of the nineteenth century, particularly in competition work at ploughing matches, this style of ploughing also required the raising of the cutting edge of the share if the 'improved' plough was to produce furrow-slices with sharp-edged tops. As with the earlier implements and techniques, this raising of the share also meant that a ribbed furrow bottom was created.

The oblique angle at which the shares of old Irish long-beamed ploughs were set, along with their flat mouldboards, produced loose, vertically standing furrow-slices. Various dimensions have been given for the sods turned. Arthur Young found that at Furness in county Kildare, ploughs penetrated the ground to a depth of nine inches, while Thompson recorded ploughing in county Meath which cut furrow-slices twelve inches wide.[18] Agriculturalists were generally critical of the shape and size of the furrow-slices produced, preferring improved ploughs which turned compact sods which lay flat on the ground.

Three people were often required to operate the old Irish long-beamed plough. A ploughman guided the plough with its handles, a driver walked in front and controlled the animals, and sometimes a man or boy leant on the plough beam, or pressed down on it with a stick, to keep it steady. Both horses and oxen were used for ploughing, although the latter tended to be confined to the farms of landlords and strong farmers.[19] The number of animals in the team seems to have varied between four and six, depending on the work carried out, and the type and condition of soil. Four animals were used for the initial breaking up of ground in winter and spring, and six for cross ploughing and ploughing summer fallows.[20]

Fallowing was a way of preparing land for grain crops. There were two distinct types. 'Green-fallowing' involved the cultivation of a root crop such as potatoes or turnips as part of the process of breaking up and cleansing land. 'Summer' (also known as 'bare' or 'naked') fallowing did not involve the cultivation of any crop, but repeated ploughings of the land during the year. Cross ploughing, which was known in many parts of Ireland as 'gorrowing' (from the Irish *gorail*), was particularly important in summer fallowing, which was the most widespread type practised in late eighteenth — and early nineteenth-century Ireland.[21] Since this fallowing most often preceded the cultivation of wheat, it is tempting to suggest a relationship between wheat-growing areas, summer fallowing, and the distribution of the long-beamed plough within Ireland. Unfortunately we do not have sufficient evidence to do more than speculate about this possible connection.

Light Ploughs

There is even less information on lighter Irish ploughs than for the long-beamed ploughs. In the early nineteenth century, however, agriculturalists recorded the use of light wooden ploughs drawn by two horses in many areas, including counties Armagh, Cavan, Cork, Derry, Donegal, Down, Kilkenny, Mayo and Sligo.[22] Despite the lack of detailed descriptions, the functions of these ploughs within Irish tillage systems can be inferred to some extent. Several accounts link the use of light ploughs with areas of light soils. The plough used in Carndonagh, county Donegal was of such a light construction that J.N. Thompson remembered, 'When I came here in 1854 . . . The plough was a light wooden implement which a boy of eighteen could fix on one shoulder and carry from townland to townland.[23] In Donegal, two-horse ploughs were used to 'prepare land for grain and sometimes for potatoes, the weight or tenacity of the soil seldom requiring a third horse'.[24] In the barony of Tyrera, county Sligo, two-horse ploughs were used to make furrow-slices five inches deep, in soils which were described as light and gravelly.[25]

Available evidence suggests that in areas of heavy clay loam, such as the good wheat-growing regions of county Kilkenny, light ploughs were employed 'in seed sowing, planting potatoes, and other easy work'.[26] One tillage practice in which the light Kilkenny plough was used in conjunction with other implements was in the Irish technique of trenching. The plough, described as being 'of the same structure' as the Kilkenny long-beamed plough but only having a beam of five feet, was said to run between the ridges to loosen the earth which was then thrown on top of the ridges with a shovel.[27]

Light ploughs in other areas, when used in the sowing process, were used like the light Kilkenny plough for covering seed. The 'common' plough of county Cork, which was said to have short, thick handles, a beam which was 'low and bending a little to the right hand, and a short pointed share without any side blade or feather', was judged to 'execute its best operations sowing seed'.[28] In 1802, Tighe described the process as it was carried out in county Kilkenny, when potatoes were being planted:

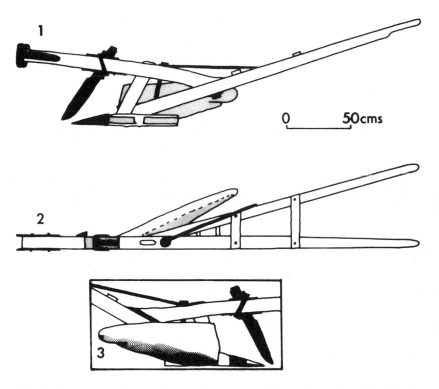

Fig. 25. A light wooden plough from Aughnish, Co. Galway, thought to date from 1780 (UFTM, L2318/1).

> The method of planting potatoes with the plough is adopted over great parts of the county; dung being spread, six furrows are usually made in a ridge, and a woman following the plough, drops the sets in each furrow about six inches asunder, the dung if long is raked over them, and each furrow is covered by the plough as it opens the one adjoining.[29]

Light ploughs were also used to cover cereal crops. Some farmers argued that covering seed with a light plough was more effective than harrowing it in, as harrowing did not cover the seed deeply enough.[30]

One surviving specimen of an eighteenth-century light Irish plough, which has some 'improved' features (Fig. 25), has been preserved in Power's Distillery in Dublin. The plough, from Aughnish, county Galway, is said to have been used to turn the first field in which barley was grown for the distillery, in 1780. It is a light wooden swing plough, with a three-sided body formed by the long handle, the beam and the breast. This construction is generally thought to be a product of

Fig. 26. A wooden plough in use near Kinvarra, Co. Galway (photograph courtesy of the National Museum of Ireland).

improved plough design. The plough also has a short beam and long handles. These 'improved' features were intended to give the ploughman more leverage, and therefore more control over the plough. The convex, curved mouldboard, the muzzle and fixed beam of the plough are also probably products of improved design. The Aughnish plough may be the forerunner of wooden swing ploughs made in south Galway and north Clare until very recently (Fig.26). These are so similar to the Aughnish plough that it seems likely that there has been a continuing tradition of plough-making in this area.

Early Developments in Plough Design

An interest in improving plough design was already evident in Ireland by the 1750's. The classic British text, *A Compleat body of husbandry*, compiled from the papers of Thomas Hale, was reprinted in Dublin in 1757, 'At the Request, and upon the Recommendations, of several Members of the Farmers' Societies in Ireland'.[31] This work described and illustrated several experimental wheel ploughs.[32] By 1767, the agriculturalist John Wynn Baker's factory near Cellbridge, county Kildare, was manufacturing eleven types of plough, including two wheel ploughs. Baker, however, had reservations about the efficiency of ploughs fitted with wheels:

> Wheel-ploughs . . . cannot be so effectual in general use, as ploughs without wheels, for this plain reason; that as the wheels are the gauge for the depth of the plough, wherever they meet with any thing which raises them, the plough consequently rises so as to plough shallow, and sometimes not to touch the surface; at other times, when the wheels sink into any declivity, the plough immediately sinks in proportion, so that the ploughing is render'd irregular by those kinds of accidents, and will continue to be so until the ploughs have been at work upon the same land for some years. Another consideration against them is, that they are in general complicated, and not a little expensive.[33]

Despite Baker's doubts, however, Arthur Young found that a few years later some landlords were importing English wheel ploughs, from Norfolk, Suffolk, and Kent.[34] One great landowner, Lord Shannon, went so far as to bring a plough and ploughman from France. This was part of a project to introduce oxen on to his estate as draught animals. His workers formed a combination against this change, until he offered higher wages to those men prepared to learn the new skills required.[35] Young also found wheel ploughs being used in the Palatine Colony at Arbella, county Kerry, while at Adair in county Limerick, the Palatines had ploughs fitted with hoppers through which seed fell as the ground was turned.[36]

By the early nineteenth century, however, the consensus among agriculturalists was that improved Scottish swing (i.e. wheelless) ploughs were best suited to most Irish conditions. In the barony of Imokilly, county Cork, in 1810, a 'good many' Scottish ploughs had been introduced, but only a few English wheel ploughs.[37]

The most famous Scottish ploughwright was James Small, who attempted to lay down scientific principles for plough design. He concentrated especially on developing the best curvature on plough mouldboards. This curve determined how well the plough turned sods over. Small argued that, 'The back of the sock [share] and the mouldboard should make one continued fair surface, without any interruption or sudden change. The twist therefor, must begin from nothing at the point of the sock, and the sock and the mouldboard must be formed by the very same rule.'[38] Small's early ploughs had wooden mouldboards, but after 1763 he began to manufacture them from iron.

Small's designs were later modified by Wilkie of Uddingston, and this developed design changed very little before the late nineteenth century. Few agriculturalists had anything good to say about 'common' Irish ploughs, when comparing them to the new Scottish implements. The Scottish ploughs (Fig. 2) required only two horses, or oxen and one ploughman to work them, and were also believed to turn sods over much more efficiently. As with so many proposed changes, agriculturalists were frustrated that Irish farmers did not immediately adopt the new ploughs. The absurdity of Irish attitudes was derided in an early nineteenth-century poem, 'The Irish Plough and the Scotch Plough':

> An Irish ploughman, with much toil and pain,
> Had worked a light Scotch plough against the grain,
> For seven long years reluctant in Fingal,
> Being asked, "pray Paddy how d'ye like it now?"
> Exclaimed, "Ochone, give me the Irish plough,
> It is the plough for Ireland after all!"
> "Why, Paddy, you're a most ungrateful rap!
> The Scotch with ease works nearly double".
> "Ah sure enough, and saves a world of trouble,
> but still the Irish is the plough for crap" —
> "Nay, Paddy, that you know is not the case".
> "Why sure enough your honor's craps increase" —
> "Then what objection can you have, you oaf?"
> "Why then, I'll tell your honor's honor why
> Before your honor laid the ould plough by,
> We always had a mighty bigger loaf".[39]

Some observers, however, recognised that there were important obstacles to the rapid introduction of the new ploughs. Hely Dutton, commenting on the use of Scottish ploughs in county Dublin, concluded that most resistance came from 'ignorance and prejudice [which] say it will not do for this country; that it kills the horses, etc., etc., etc.'[40] However, he also believed that those who disliked the new ploughs might have been working with ones which were not properly set.[41] The adjustment of the coulter, especially, was important if the ploughs were to be used to their full advantage. New skills, therefore, were required, not only of ploughmen but of ploughmakers, and local blacksmiths.[42]

Other problems identified were that the new ploughs were more expensive than the old,[43] and also that hired ploughmen would be reluctant to do without the help of the other men required to work the older, heavy ploughs, as these were often members of the same family:

> Old ploughmen are very much averse to the Scotch plough and they therefore throw every obstacle in its way; . . . perhaps . . . [the ploughman's] son has been the driver upon the old system, in which case little is to be expected from the father's exertions, as the greater the degree of perfection, to which he brings the Scotch plough, the farther will his son be from employment.[44]

One solution suggested to this problem was that farmers should pay their ploughmen, first their own wages, but also half of the driver's wages.[45] This clearly illustrates how determined agriculturalists were that the new Scottish ploughs should be used. Even the most enthusiastic supporters of the new ploughs, however, agreed that in some cases they did not perform as well as 'common' Irish ploughs. Richard Thompson, for example, admitted, 'I am, and have been, since I first saw the Scotch plough, a very great advocate for its being generally adopted, yet I confess that in this country there are many soils in which it will not answer for the summer's gorrowing [i.e. cross-ploughing summer fallows]'.[46] As late as 1819, William Shaw Mason reported that in county Kilkenny, although

> Some of the more respectable farmers have introduced Scotch ploughs . . . they are not suited to wet grounds, which are so stiff and adhesive, that the sod requires to be raised as much as possible, in order that the air may get through it, to dry and pulverise it. The Scotch plough is thought to lay the sod too flat, while the Irish plough raises it, and breaks it more; . . . [while] the superiority of the Scotch plough in dry grounds is fully acknowledged, a dislike to it . . . in the former instance may have some foundation in reason. However when, in the advancing improvement of agriculture, those grounds were drained . . . pulverisation will be effected by the Scotch as well as the Irish plough.[47]

A reluctance to use Scottish ploughs, therefore, was not simply due to unreasonable conservatism. New skills had to be learned by ploughmen and ploughmakers, the initial expense was greater, ploughmen were concerned at possible job losses arising from the new technology, and finally, the new ploughs only performed at their best on well-drained ground, and in systems of agriculture where summer fallowing was not practised. Hely Dutton recognised how easily new implements and techniques could be discredited:

> We may continue to import Scotch swing, and English wheel-ploughs, and other implements of utility and whim, but unless we import along

with them the best practices of each country, and steadily pursue them in opposition to the old school stewards, it will only serve to bring them into disrepute with those who are but too ready to catch at any opportunity to decry practices they do not understand.[48]

There were several attempts to improve Irish ploughs, but the adoption of Scottish swing ploughs did continue. The enthusiasm of farming societies, and their success in persuading Irish farmers to start using the new ploughs, can be traced through accounts of ploughing matches reported in the *Munster farmer's journal*.

In 1812, the Cork Institution decided that at its ploughing matches for the next year there should be two classes of premium: one for the ploughmen of 'gentlemen and farmers', and the other for 'working farmers'. Irish ploughs could only be used by the second category.[49] Ploughs of Scottish design, some made locally, were given as prizes. The same year, the Farming Society of Ireland employed a champion ploughman, William Kippie, to instruct farmers in the use of Scottish ploughs. Kippie would 'attend any farmer who wishes to employ him at 2s. 6d. per day with his diet and lodging'.[50] In a ploughing match at Fermoy, County Cork, in 1813, separate prizes were given for Irish and Scottish ploughs,[51] but by 1816 the Scottish ploughs were becoming dominant. At a ploughing match at Kinalea, county Cork, on 13th March, twenty-one ploughs were entered, twenty of them of Scottish design. The report concluded that

> The rapid improvements in ploughing . . . and the very general adoption of Scotch ploughs which have been introduced into this district within the last few years, afforded a most satisfactory evidence of the utility of ploughing matches — and the great object for which they were instituted having been achieved, the members have resolved to limit the number of such contests, to one annually.[52]

Even those county Cork farmers who had not actually bought a Scottish plough were said to have improved their own implements, and learned to plough with two horses and without a driver.[53]

Small, locally designed wooden ploughs survived, however, sometimes as subsidiary implements to the metal Scottish ploughs. In 1837, for example, Binns found that in the west of county Waterford, 'Previously to sowing grain, the land is ploughed with the Scotch plough, and well harrowed; then a small wooden plough without iron in the mouldboard is used for the purpose of ploughing in the seed. Under this process, the ground, not being left smooth, is less liable to become hard and stiff'.[54]

Fig. 27. A metal swing plough in use in Ulster, early this century (WAG 291).

Nineteenth-Century Improved Plough Types

By the mid-nineteenth century, all-metal Scottish swing ploughs (Fig. 27) had become the most common improved plough type in most parts of Ireland. The catalogue for the Irish Industrial Exhibition of 1853 described the swing plough as 'almost universally employed'.[55] Makers of improved agricultural implements multiplied throughout the county during the early part of the century. In 1812, in county Antrim, for example, it was reported that, 'Within the last ten years a manifest improvement has taken place in the implements of husbandry . . . they are rapidly spreading over the county; manufacturers of them being established in many places'.[56] The cast metal mouldboards and shares of the improved Scottish swing ploughs were suited to large-scale industrial production, and foundries began to produce standardised ploughs and plough parts. The best-known foundry in the north of Ireland was set up by Robert Gray in Belfast in 1840, initially as a subsidiary of its parent company in Glasgow. Other famous foundries producing swing ploughs around the same period as Grays were Paul and Vincent of Blackhall-Place, Dublin and W. & J. Ritchie of Ardee, county Louth. Foundries also sold plough parts to blacksmiths who constructed implements from them. In the later nineteenth century, for example, blacksmiths in north Antrim obtained sole plates, mouldboards, and roughly blocked out shares and parts of plough bodies from the Coleraine firms of Kennedy and Son, and Moore and Sons.[57]

Making a swing plough to the correct proportions, and adjusting the working parts to the right position, were tests of skill for both foundries and rural blacksmiths. Ploughmakers needed to have an intimate knowledge of the relationship between various parts of the plough when the implement was in use. A common complaint in the early nineteenth century was that both kinds of knowledge were sadly lacking in Ireland. In 1839, John Sproule wrote that 'The adaption of the improved Scotch plough for the execution of the work required, shows that the most minute acquaintance with practical mechanics was necessary for its construction. This knowledge is, however, rarely possessed by the farmer, nor even by the greater number of those who now manufacture the implement'.[58] A similar complaint was made in *Purdon's practical farmer* in 1863: 'In some cases, a modern form is aimed at; but the proper principles upon which the plough should be constructed being unknown to the makers, the implements which they turn out are generally deficient, performing the work in an indifferent manner'.[59]

Before a swing plough was used, the coulter and share, or 'irons', were set to perform whatever work was required. To do this properly, blacksmiths required not only a knowledge of ploughmaking, but of ploughing techniques. Setting the coulter required less skill than adjusting the share, so many farmers were able to do this themselves. However, the best position for the coulter was a matter of controversy amongst nineteenth-century agriculturalists. *Purdon's practical farmer* commented that, 'Simple though the form and duties of the coulter may be, there is no member of the plough whereof such a variety of opinions exist as to its position'.[60] If the point of the coulter was set too close to the share, or to the land side of the plough, it gave a very sharp top to the furrow slice. This could be a disadvantage, even in ploughing matches, as the sharp crests of the furrow slices were liable to crumble, so making the distance between them difficult to measure.[61]

There were two main types of furrow slices turned by Scottish swing ploughs. These were known as rectangular, and high-crested, work (Figs. 28a and 28b). For rectangular work, the cutting edge of the 'wing' or 'feather' of the share was flattened by the blacksmith and the plough fitted with a concave-shaped mouldboard. The development of the concave or straight-lined mouldboard for rectangular work is

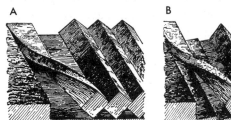

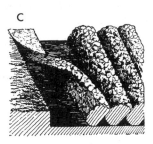

Fig. 28. Three types of furrow-slice produced during ploughing: A. Rectangular work. B. High-crested work. C. Broken work (Wright, R. P. (1908), vol. 9, p.252).

attributed to James Small. High-crested work required a much finer adjustment to the share and the use of a convex shaped moulboard. James Wilkie is credited with introducing the convex mouldboard into improved Scottish swing plough design. Shares set for high-crested work produced a 'trapezoidal' furrow slice. The ploughing style seems to have been introduced into Ireland in the mid-nineteenth century, *Purdon's practical farmer* noted that, 'Of late years . . . a form of furrow has been introduced which presents a sharper and high crest, the furrow slice being cut at a very acute angle'.[62] To produce this, the cutting edge of the share had to be raised to a precise angle and shape. The share was shaped on the blacksmith's anvil, using a hammer and tongs. To ensure that the shape was correct, some blacksmiths tested their work against templates which gave the exact curvature.[63]

The rectangular and high-crested furrow slices turned by swing ploughs provided an excellent seed bed for grain. When seed was sown broadcast over the area ploughed, it tended to fall into the furrows, and when harrowed, the tops of the furrow slices were knocked off to cover the seed. In carefully ploughed fields, the seed grew in rows not unlike those produced when a seed-drill was used. The technique was encouraged, even at the latter end of the nineteenth century: 'In ploughing lea land, that is intended for a grain crop to be sown broadcast, either the rectangular, or the high-crested furrow, is best suited, as in this system the seed falls between the furrow slices. In good ploughing there is sufficient earth on the crests of the furrow slices to cover the seed that is sown open'.[64]

Swing ploughs remained popular in many parts of Ireland until well into the present century. This was especially true on small farms in hilly areas. The main reason seems to have been that identified almost two centuries before by John Wynn Baker, quoted earlier in this chapter. The ploughman could adjust the draught (or strain on the horses) of a swing plough at will. When the handles were lifted upwards, the share pointed downwards, cutting a deeper furrow slice and so increasing the draught. The width of the furrow slices could also be changed by pressure on the handles. Going uphill, a narrower furrow slice, which required less draught, could be cut. Both of these possible adjustments meant that the ploughman could ease, or increase, the strain on the horses without having to stop and adjust the plough. On ploughs with wheels, however, the depth and width of the furrow slice cut were determined by setting the wheels at the front of the implement before ploughing was commenced. In order to change the draught, the ploughman had to stop the plough and adjust the wheels. This made them less attractive than swing ploughs, not only on hilly land, but also on rough, stony ground. Swing ploughs could be eased over large stones by raising and lowering their handles as they were driven along, whereas wheel ploughs, being fixed, would bounce off large stones, and sometimes miss a piece of ground.[65]

Wheel ploughs had important advantages over swing ploughs, however. On level ground, once the wheels were set, the ploughman had simply to guide his horses to produce uniform furrow slices. Much less labour, and less skill, were required when these ploughs were used. It is significant that the biggest manufacturers of wheel ploughs in Ireland were located in Wexford, near the major tillage areas of Leinster. Wheel ploughs were well suited to work on large, well-drained, lowland farms. The ploughs manufactured in Wexford (Fig. 29) were mostly based on English prototypes, but by the end of the nineteenth century, light American ploughs had also become popular, especially those manufactured by the Oliver Chilled Plough Works, Indiana. The American ploughs had revolutionary 'digging' mouldboards designed to execute a wide broken-back furrow slice (Fig. 28c). This was particularly suited to the drill cultivation of crops, as the soil was pulverised to a sufficient depth to allow the raised drill cultivation of green or root crops such as potatoes, and to a fine enough tilth for the horse-drawn seed drill cultivation of grain crops.[66]

The short, wide-set, concave digging board made of chilled steel was a design which became increasingly popular with farmers and was adopted and adapted by ploughmakers in Ireland. The digging plough, often referred to as the chill plough because of the chilled steel used in its construction, was enthusiastically proclaimed by one Irish farmer in a farming journal of 1910 to be 'the plough that suits the farmer best, as she is a far lighter draught, and therefore best adapted for turning

Fig. 29. A wheel-plough, with a 'semi-digging' board, manufactured by Pierce of Wexford (Pierce catalogue (195-?), (p.11).

and breaking a big slice'.[67] Even as late as the 1950's, agricultural machinery makers Pierce of Wexford identified their A.R. two-wheel semi-digger horse plough as 'one of our most popular models.'[68] (Fig. 29).

Wheel ploughs of all kinds became increasingly common in Ireland during the later nineteenth century. This can be related, in part, to the drainage and consolidation of farms during this period, but at a technical level, it was also due to the development of ploughs with interchangeable, standard parts. The English firm, Ransomes of Ipswich, developed a 'universal body' plough in 1808.[69] Throughout the nineteenth century, other foundries also began to manufacture ploughs with interchangeable parts. In Ireland, it was again the large Wexford companies, Pierce's and the Wexford Engineering Works, which were most involved. By 1950, some of the ploughs manufactured in the Wexford Engineering Works could be fitted with parts made by the British firms, Ransomes, Howard, Hornsby, and Sellar, and the American firm of Oliver.[70] These parts could be bought and fitted by farmers without the help of blacksmiths, and the use of the standardised ploughs was further increased by the closing of many small country forges during the present century. As with so many aspects of farming technology, however, developments overlapped one another. One county Down farmer, interviewed recently, remembers that at a Royal Ulster Agricultural Show in the late 1930's, Ferguson tractors attracted less attention than a new horse-drawn wheel plough, manufactured by Ransomes.[71]

Plough Teams and Ploughing Techniques

Attempts to use oxen systematically on large Irish farms continued at least until the end of the First World War,[72] but most Irish farmers used a cow or a bullock for draught only if they could not make a plough team up in any other way. During the nineteenth century, the typical plough team in Ireland increasingly became two horses, harnessed abreast. In the later part of the century horses of the Irish draught type became common in south-eastern Ireland, while in the north-east, the Clydesdale breed became popular. Many farmers, however, continued to use smaller horses, often bred from small hill ponies, which in turn were at least partly descended from the garrans (Irish: *gearráin*) of Gaelic Ireland.[73]

Teams were harnessed in various ways. One of the most notorious practices was to attach the plough to horses' tails (Fig. 30). This seems to have continued in Erris, county Mayo, until 1841.[74] Its inefficiency might have been excusable if no alternative was available, but in fact cheap, efficient and humane harness could be made by farmers from twisted straw (Irish: *súgán*), and withies (Irish: *gaid*). Collars and backbands were made from plaited straw rope (Fig. 21), while trace ropes were made by twisting strips of bog fir. Straw harness was recommended by nineteenth-century agriculturalists for breaking in young horses, and even as an alternative to the heavy yokes used for harnessing oxen.[75] The use of straw harness persisted well into the present century, and some older farmers can still make collars. Leather harness became more standardised during the nineteenth century. Two main types of collar became popular with the Irish farmers. One was the open-ended collar, which was slipped around the horse's neck and strapped at the top. The other was the closed collar, which was put over the horse's head upside down, and then turned on the horse's neck into the correct upright position.

Ploughs and plough teams could be used in many different ways. The dimensions and shapes of some cultivation ridges have been discussed in previous chapters. The techniques of ploughing by which ridges were built up, varied very widely, however. Ploughs were often used in conjunction with spades during the primary tillage stages in ridge making. One method for the construction of potato ridges, remembered by Mr. John McElroy, a small hill farmer from county Monaghan, was practised on steeply sloping ground with a chill or digging plough and spade, when only one horse was available for ploughing. The average width of a ridge was said to be six furrows,

Fig. 30. An anti-Irish caricature, lampooning the practice of ploughing by the tail. The drawing is almost certainly inaccurate as the nearest horse is wearing a bridle, but no reins, and also has a draught collar, which would have been superfluous as there are no trace-chains (Dubois, E. (1808), p.82).

with half the ridge ploughed and half turned with the spade. As the uphill draught of the plough was excessive for one horse, ploughing could only be done downhill. Consequently the part of the ridge normally turned by the uphill passage of the plough had to be turned with the spade. Manure was spread on top of the ridges and the potatoes set on top of the manure and then covered with earth from the trenches between the ridges. The use of a spade and chill plough for high narrow ridge construction is a striking example of how small farmers, constrained by limited resources and geographical location, adapted and combined old and improved technologies to work within established cultivation systems.

When ploughs alone were used for ridge construction, very narrow strips of untilled ground were sometimes left simply as an accidental by-product of ploughing. In the techniques most praised by commentators, a small furrow was made along the central line of each ridge, and the next furrow slices were turned over this, to meet in the middle. If this was not done, a thin strip of unturned ground could be left along the middle of the ridge. However, strips were often deliberately left during the initial ploughs. In 1832, the diarist Amhlaiobh Ó Súilleabháin listed some Irish words related to ploughing, which included a word for these untilled strips: '*Fód*, a sod: *fódfhughála*, the unploughed sod between the first two sods: *caolfhód*, the last sod ploughed in a ridge, or set (*seathach*), which leaves the *grianán*, or hollow, exposed to the sun.'[76] In the 1770s, Arthur Young noted several instances of plough ridges constructed around fairly broad untilled central sections. At Annsgrove, for example, he found that Plough teams could plough an acre of land in a day, 'by means of leaving a great space in the middle of each land, where they begin by lapping the sods to meet'[77] In the barony of Forth, county Wexford, one third of the ridge was left unploughed in the middle.[78]

The practice of making lazy-beds, or constructing ridges over untilled strips of land has already been described in connection with spade work. In many parts of Ireland, ploughs were used to mark out lazy-beds by turning only one furrow slice at each side of the ridge. In 1802, for example, Coote found that farmers in county Cavan 'mark out the ridges with the plough, spread the dung on the grass, on which is laid the seed, and throw up the earth from the trench'.[79] Similarly in county Cork, Townsend claimed that, 'Formerly, it was usual in many places to plough or dig the trenches only, leaving the beds whole, on which, after the dung was laid on, the potatoes were placed, and covered from the trenches with the shovel'.[80] Lazy-beds were also marked out with ploughs in at least one large-scale scheme of reclaiming mountain land in Queen's county (Offaly) during the 1840s,[81] and the practice has continued into the present century. Evans describes a technique where a paring plough, fitted with a wide share, was used. This could turn a broad furrow slice without breaking the sod.[82] On the Dingle peninsula, county Kerry, potatoes are still planted on lazy-beds (Irish: *iomairí*) constructed using an all-metal swing plough (Fig. 31), spade and shovel. Nowadays the ploughs are often pulled by tractors, but farmers say that they use the same technique as when horses were used.

In one recorded example,[83] the farmer first ploughed rows of furrow slices across the field to mark out the edges of the ridges and furrows. The furrow slices were turned so that they fell in towards the strips of ground on which the ridges were to be built up. After this task was completed, a one-sided spade or loy (Irish: *láighe*) was used to cut out the narrow strips of sod which had been left in the middle of the furrows between ridges. These sods were placed along the middle of each untilled ridge to form a central spine. The final, or trenching stage

Fig. 31. The edges of cultivation ridges (*iomairí*) being marked out using a swing plough, near Slea head, Co. Kerry (photograph courtesy of *An Clodhanna Teoranta*).

in making the ridges again involved using the plough, and also a long-handled shovel. The mouldboard of the plough was removed, and the bottom of each ridge furrow, or trench, was then ploughed to break up the soil. The loose soil was shovelled on to the top of adjacent ridges. Animal manure, and in this coastal area seaweed, were also added. After the potatoes had been planted, more loose earth was shovelled from the trenches to cover the seed. When complete, the ridges were just over one metre broad, and the furrows about half this. The ridges were about a quarter metre in height. These potato ridges have very steep sides, but it seems likely that similar techniques were also used when oats were grown.[84] In lazy-bed construction, the use of an improved swing plough, along with a loy and long-handled shovel, and in modern times the replacement of horses by tractor for draught power, are further examples of small farmers adapting and combining new and older technologies to work together within long-established cultivation systems.

During the nineteenth century, with the spread of improved plough types and increased land drainage, ploughing practices became more

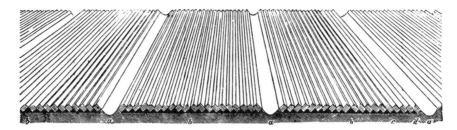

Fig. 32. A standard method of ploughing flats in the late nineteenth and early twentieth centuries. a-b Furrow slices lying from left to right b-a Furrow slices lying from right to left c-b Crowns of ridges (and line of 'scores') d Open-furrows or 'hint-ends' (Stephens (1908), vol.1, p.394).

standardised, especially on large lowland farms. Broad, low ridges know as 'flats' were built up, using the most common technique known as 'gathering' and 'casting' (Fig. 32). Nineteenth-century farming texts give us clear accounts of this method, which is still used in modern horse-ploughing matches. First, the middle line of the ridge was marked out or 'scored'. The plough was pulled up one side of this line and down the other, in a clockwise direction. This was called 'gathering' and produced two series of parallel furrows. One line of furrows lay inclined to the right, the other to the left, and they met in the middle or 'crown'. 'Casting' was done when two flats were ploughed at the same time. This involved ploughing in an anti-clockwise direction down the left-hand side of one flat and then along the right-hand side of the adjoining flat. Casting was done prior to joining two flats. Where two flats met, an 'open furrow' was formed by the slices turning away from one another. The open furrow was narrowed, and deepened, by smaller furrow slices being ploughed along each side. These were known as 'hint ends'[85] and assisted in the drainage of surface water.

Ploughing techniques had obvious implications for other aspects of cultivation, including how seed was sown, and how crops were harvested. In later chapters some of these connections will be discussed in more detail. These will further emphasise how important it is to consider any implement and the way it is used, not only in itself, but also in relation to the whole system of cultivation of which it is a part.

NOTES

1. Doyle Martin, *A cyclopaedia of practical husbandry*, rev. ed. by W.H. Rham (London, 1844), p.437.

2. Watson, M., 'Common Irish plough types and tillage practices', *Tools and tillage*, vol. 5:2 (Copenhagen, 1985), p.85.

3. *Ibid.*

4. Rawson, T., *Statistical survey of the county of Kildare* (Dublin, 1807), p.7.

5. Tighe, W., *Statistical observations relative to the county of Kilkenny* (Dublin, 1802), p.293

6. *The Munster farmer's magazine*, vol. 1 (Cork, 1812), pp. 43-46; Watson, M., *op. cit.*, p.87.

7. Thompson, R., *Statistical survey of the county of Meath* (Dublin, 1802), p.109.

8. Watson, M., *op. cit.*, p.87.

9. Thompson, R., *op. cit.*, p.109.

10. *Ibid.*, p.110

11. Sampson, G.V., *A memoir, explanatory of the chart and survey of the county of Londonderry* (Dublin, 1802), p.185.

12. Watson, M., *op. cit.*, p.89.

13. Marshall, G., 'The "Rotherham" plough', *Tools and tillage*, vol. 4:3 (Copenhagen, 1982), p.134.

14. Tighe, W., *op. cit.*, p.293.

15. Thompson, R., *op. cit.*, p.110.

16. Watson, M., *op. cit.*, p.90.

17. Dutton, Hely, *Observations on Mr. Archer's statistical survey of the county of Dublin* (Dublin, 1802), p.33.

18. Watson, M., *op. cit.*, p.91.

19. Bell, J., 'The use of oxen on Irish farms since the eighteenth century', *Ulster folklife*, vol. 29 (Holywood, 1983); Lucas, A.T., 'Irish ploughing practices, part 3,' *Tools and tillage*, vol. 2:3 (Copenhagen, 1974).

20. *Ibid.*

21. Watson, M., *op. cit.*, p.92.

22. *Ibid.*

23. Harkin, Maura and McCarroll, Sheila, *Carndonagh* (Dublin, 1984), p.20.

24. McParlan, James, *Statistical survey of the county of Donegal* (Dublin, 1802), p.28.

25. McParlan, James, *Statistical survey of the county of Sligo* (Dublin, 1802), pp.114-117.

26. Tighe, W., *op. cit.*, p.293.

27. *Ibid*, pp.183-184.

28. Townsend, H., *Statistical survey of the county of Cork* (Dublin, 1810), pp. 190-191.

29. Tighe, W., *op. cit.*, p.218.

30. *Ibid.*, p.198.

31. Hale, Thomas, *A compleat body of husbandry*, vol. 2 (Dublin, 1757), title page.

32. *Ibid.*, pp. 60-102.

33. Baker, John Wynn, *A short description and list, with the prices of the instruments of husbandry, made in the factory at Laughlinstown, near Cellbridge, county Kildare* (Dublin, 1767), pp. 10-11.

34. Young, Arthur, *A tour in Ireland*, vol. 2 (Dublin, 1780), p.32.

35. *Ibid.*, p.56.

36. *Ibid.*, pp.123 and 138.

37. Townsend, H., *op. cit.*, p.613.

38. Fussell, G.E., *The farmer's tools, 1500-1900* (London, 1952), p.49.

39. *The Dublin penny journal*, vol. 2 (Dublin, 1833), p.150.

40. Dutton, Hely, *op. cit.*, p.33.

41. *Ibid.*

42. Dubordieu, John, *Statistical survey of the county of Down* (Dublin, 1802), p.50.

43. *Ibid.*

44. Thompson, R., *op. cit.*, p.114.

45. *Ibid.*

46. *Ibid.*, p.113.

47. Mason, William Shaw, *A statistical account and parochial survey of Ireland*, vol. 3 (Dublin, 1819), p.633.

48. Dutton, Hely, *Statistical survey of the county of Clare* (Dublin, 1808), p.37.

49. *The Munster farmer's magazine* vol. 1 (Cork, 1812), p.342.

50. *Ibid.*, p.85.

51. *The Munster farmer's magazine*, vol. 2 (Cork, 1813), pp. 85-87.

52. *The Munster farmer's magazine* vol.4 (Cork, 1816), pp. 152-153.

53. *Ibid.*, p.78.

54. Binns, J., *Miseries and beauties of Ireland*, vol. 2 (London, 1837), p.296.

55. Sproule, John (ed.), *The Irish industrial exhibition of 1853: a detailed catalogue of its contents* (Dublin, 1854), p.202.

56. Dubordieu, John, *Statistical survey of the county of Antrim* (Dublin, 1812), pp. 151-152.

57. Watson, M., 'North Antrim swing ploughs: their construction and use', *Ulster folklife*, vol. 28 (Holywood, 1982).

58. Sproule, John, *A treatise on agriculture* (Dublin, 1839), p.51.

59. *Purdon's practical farmer* (Dublin, 1863), p.147.

60. *Ibid.*, p.155.

61. Watson, M., *op. cit.* (1982), p.18.

62. Purdon, *op. cit.*, p.153.

63. Watson, M., *op.cit.* (1982), p.19.

64. Irish national schools, *Introduction to practical farming* (Dublin, 1898), p.10.

65. Watson, M., *op. cit.* (1982), pp. 20-21.

66. Irish national schools, *op. cit.*, pp.10-11.

67. *The Ark*, no. 3 (Belfast, February 1910), p.19.

68. Philip Pierce and Company Limited, Wexford, *Catalogue* (Wexford, 195-?), p.11.

69. Partridge, Michael, *op. cit.*, p.42.

70. Wexford Engineering Company, *Catalogue* (Wexford, 1950).

71. Interview with J.A. Weir, Ballyroney, Co. Down (Ulster Folk and Transport Muscum Tape C78.54).

72. Bell, J., *op. cit.*, p.26.

73. Watson, M., 'Cushendall hill ponies', *Ulster folklife*, vol. 26 (Holywood, 1980); Lucas, A.T., 'Irish ploughing practices, part 2', *Tools and tillage*, vol. 2:2 (Copenhagen, 1973).

74. Lucas, A.T., *ibid.*, p.78.

75. Doyle, Martin, *op. cit.*, p.412.

76. Ó Súilleabháin, Amhlaiobh, *Cinnlae Amhlaiobh Úi Shúilleabháin*,vol. 3, trans. M. McGrath (London: Irish Texts Society, 1930 (1936)), p.143.

77. Young, Arthur, *op. cit.*, vol. 2, p.12.

78. *Ibid*, vol. 1, p.111.

79. Coote, C., *Statistical survey of the county of Cavan* (Dublin, 1802), p.73.

80. Townsend, H., *op. cit.*, p.195.

81. Trench, W.S., *Realities of Irish Life* (London, 1868), p.97.

82. Evans, E.E., *Irish folkways* (London, 1957) (1967)), p.144.

83. Bell, J., 'A contribution to the study of cultivation ridges in Ireland', *Journal of the Royal Society of Antiquaries of Ireland*, vol. 14 (Dublin, 1984), p.84.

84. *Ibid.*, p.88.

85. Stephens, Henry, *The book of the farm*, vol. 1, 23rd ed., (Edinburgh, 1871), p.97.

CHAPTER 5

Harrows, Rollers and Techniques of Sowing Seed

Harrows and rollers can be used at several stages of the annual cycle of crop cultivation. During the last two centuries, however, they have been most frequently used just before, and just after, seed was sown. This chapter will concentrate on the processes involved during planting crops. The special case of planting potatoes will be dealt with in the chapter dealing specifically with that crop.

Harrows

Harrows were used for breaking up soil and weeding it just before sowing, but also for mixing soil with newly sown seed, so burying it. Other implements could be used for the same tasks. Some of these, for example grubbers and horse-hoes, were developed for use in 'improved' systems of cultivation. On very small farms on the other hand, lumps of earth were often broken up using wood mallets known as mells, or in the south-west of Ireland, heavy hand-hoes or mattocks, known as graffans (Irish: *grafáin*). Seed could be covered using a wooden hand-rake, or by simply dragging a bush across the ground. This last technique was so successful that eighteenth-century agriculturalists recommended an 'improved bush harrow' made by twisting pieces of white thorn around an old farm gate.[1] Most harrows can be easily distinguished from all these other implements by their construction. Standard English terms can be applied to many of their parts (Fig. 33).

Harrows have probably been used in Ireland for at least one thousand years. Although little archaeological evidence has been published, even very early texts contain terms which have been interpreted by scholars as referring to harrows. The eighth-century *Críth Gablach*, for example, lists '*da chapal/i/do foirsti/u/d*' among the possessions characteristic of one of the noble grades of early Irish society. One suggested translation of this phrase is 'two horses for harrowing'.[2] References to harrowing have also been found in Norman estate records. At the Manor of Cloncurry, county Kildare, in 1304 A.D., the cost of harrowing an acre of either

90

Fig. 33. Wooden-framed harrows of a type used in the late nineteenth and early twentieth centuries. The main parts are: A. Leaf B. Bull C. Tine or pin (UFTM, L6/11).

Fig. 34. Oxen pulling harrows, near Hillsborough, Co. Down, in 1783 (detail from a print by W. Hincks).

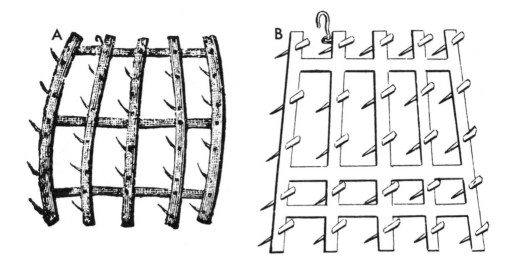

Fig. 35. A. Common harrow from Co. Meath (Thompson, R. (1802), p.116). B.
Common harrow from Co. Kilkenny (Tighe, W. (1802), p.300).

wheat or corn was given as three pence. O'Loan suggests that these
harrows were probably similar to those illustrated in the Duc de Berry's
Book of Hours (1360), constructed with wooden frames and wooden
tines.[3]

Sporadic references to harrowing continue to appear in fifteenth-and
sixteenth-century Irish writings, but it is only in the eighteenth century
that physical descriptions, and illustrations, become detailed (Fig. 34).
Both 'improved' and 'common' harrows were described, although the
latter were often only discussed in terms of how they might be improved.
As late as 1830, harrows claimed to be in common use among poorer
Irish farmers were dismissed as 'villainous'.[4] In fact, as with ploughs,
harrows described as 'common' varied quite considerably from one part
of Ireland to another, differences even being noted within the same
county. In many areas, for example, observers distinguished 'light' and
'heavy' harrows. In county Meath, in 1802, large harrows were used.
These had outer bulls which curved in towards a straight central bull
(Fig. 35a).[5] In county Kilkenny in the same year, however, the harrows
described as 'common' had straight bulls (Fig. 35b).[6] Both the Kilkenny
and Meath implements have five bulls. In Tyrone, however, the harrows
recorded as in general use in 1802 had only four bulls.[7] The number of
tines or 'pins' on common harrows also varied. In county Kilkenny the
number cited was twenty-five, in county Meath twenty-four, in
Monaghan twenty-one, and in Tyrone, twenty.[8]

Improved Harrows

Experiments to improve harrows were well under way by the mid-eighteenth century. One early aim was to produce stronger implements. In 1769, John Wynn Baker produced harrows in his factory in county Kildare which were riveted on each side of the 'pin' holes. This was intended to prevent the bulls 'being split, in driving the pins by a careless hand'.[9] Improvers in England also worked at increasing the strength of the implements, especially the great harrows or brakes used, like the heavy 'common' Irish harrows, to break up ground after ploughing. In 1757, for example, Thomas Hale recommended that the wooden parts of harrows should be made from ash and that 'a good workman should be employed; for if it be not framed very well together, it will tear itself to pieces presently in the working'.[10]

A frequent criticism of 'common' Irish harrows was that when they were pulled through the ground, rows of tines followed one another along a single track instead of each breaking up a separate piece of ground. Discussion also centred on the shape of the tines and the angle at which they should go into the ground. The best weight for harrows was also debated. Tighe's discussion of the 'common' harrows of county Kilkenny raises most of these points:

> The tines being put in without system, often follow each other in the same track, leaving intervals of six or seven inches unstirred; their points incline generally too much towards the fore parts of the harrow, and their sides are turned to the front instead of their angles: by these circumstances the draft becomes difficult and the harrow is soon clogged with soil . . . Add to this that being made in one frame it is too heavy for the driver to lift up conveniently.[11]

The claim that on 'common' harrows rows of tines simply followed one another does not seem altogether justified. Some details of their construction suggest a deliberate attempt to prevent this. Tines on the curved bulls of the early nineteenth-century Meath harrow (Fig. 35a) and on the county Kilkenny harrow, where the bulls converged towards the front of the implement (Fig. 35b), would have been unlikely to follow exactly in one another's tracks. Another attempt at a solution to the problem is suggested by the positioning of the hooks to which draught chains were attached on both the Meath and Kilkenny implements. The hooks were fitted to bulls left of centre of the harrow frames. This again would have prevented rows of tines following directly behind each other. This placing of the draught books would also have partially answered Tighe's criticism that tines were pulled through the

Fig 1

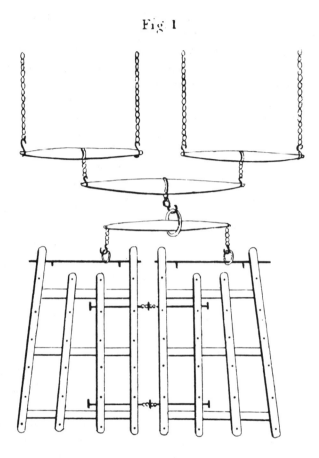

Fig. 36. An improved set of harrows from Co. Kilkenny, in 1802. The swingle tree to which the leaves are joined is attached to the other trees by a large loose ring, which allowed the harrows to 'hustle' from side to side (Tighe, W. (1802), p.300).

ground flat side on. In these arrangements, tines which were square or rectangular in section would have met the soil edge on.

It is also doubtful whether any of the 'common' harrows described by early nineteenth-century writers would have been pulled smoothly enough through the ground for the tines to follow one another in firm, straight tracks. Even some modern horse-drawn harrows are linked to the harness trace chains loosely enough to allow the implements to bounce around as they are pulled. Some early 'improved' harrows were deliberately constructed to ensure that this happened. In both counties Kilkenny and Clare, experimenters produced implements which moved with a 'hustling latitudinal [motion] . . . so desirable in rough ground'.[12]

The county Kilkenny harrow was attached to a swingle tree by a loose ring to allow this sideways movement (Fig. 36).

The concern that no two harrow tines should follow the same track can probably be related to the use of high, narrow cultivation ridges. On these ridges, cross-harrowing would have been limited. Where harrowing was carried out in only one direction, however, there was a danger that the effects of the implement would be undone by the horses' feet compacting the earth again, as they walked up and down:

> Let the husbandman . . . be cautious that he do not expect too much from [harrowing] . . ., or depend upon it in cases where he should have recourse to more powerful methods of breaking the ground. This is a caution the more necessary, because it is an error that present farmers very frequently run into . . . When they have thrown in their seed they go over it with a harrow, and being sensible that the clods of earth must be broken, . . . When they see once or twice harrowing does not effect it, they go over the ground again and again, till the feet of the horses have trod the soil in a hardness that is very unfit for the growth of anything.[13]

By 1830 the 'rhomboidal' frames of 'Scotch' harrows were widely recommended as the best to ensure that tines broke up the entire surface of the ground (Fig. 37). Martin Doyle urged his readers to 'observe the tracks in which the pins move — all separately and distinctly, and at equal distances from each other, loosening every particle of the ridge over which . . . [the harrow] moves'.[14]

The criticism of the county Kilkenny common harrows, that their tines were inclined too much towards the front of the implement, has to be balanced against the advice given by early nineteenth-century English writers, that tines on heavy harrows should have an incline towards the front.[15] Harrows with tines arranged in a perpendicular position were thought to be less effective in breaking up clods of earth. A forward tilt allowed the tines to take a better hold on the ground as the harrows were pulled. Many early twentieth-century harrows had tines shaped, in the words of one county Antrim blacksmith, 'like a plough coulter'. A forward tilt was not advised for tines on lighter harrows, however, since it was argued that this made it more likely that they would become clogged with weeds.

By 1767, John Wynn Baker's factory at Cellbridge, county Kildare, was manufacturing harrows made in several sections or 'leaves', hinged together. Baker pointed out that hinged harrows had an important advantage: 'The harrow, in general use in this Kingdom, is too often ineffectual in its operation, by its being made only in one frame, but by mine being made in two frames, united together by what I call coupling

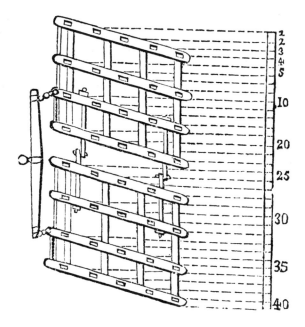

Fig. 37. Rhomboidal Scotch harrows, approved by agriculturalists because, as the diagram shows, no two tines followed the same track through the ground (Doyle, M. (1844), p.272).

bolts, they lie close to the ground, even in irregular places'.[16] This advantage was especially important where plough ridges were narrow and pronounced. In county Galway in 1824, the double harrow was recommended as it 'lies better each side of a ridge'.[17]

The other main advantage claimed for hinged harrows was that each section could be easily lifted. This has remained an important consideration in harrow design, since even modern horse-drawn implements become clogged when in use, and must be lifted from the ground for cleaning.[18] Many of these harrows have a handle attached to each leaf so that the operator can lift that section, while continuing to walk behind the harrows and drive the horses. The handles also allow pressure to be increased when the implement is crossing a high patch of ground. In this way soil can be collected and carried forward, and then shaken off in a hollow place. Most experiments in harrow design seem to have used hinged constructions. Later nineteenth-century evidence suggests an increasing use of hinged harrows throughout Ireland.

In discussions on the improvement of harrows there was a lot of disagreement. For example, some county Meath farmers in 1802 argued that making harrows in several sections 'lessens the specific gravity; not more than half the weight of the harrow being in any one point at the same time, and prevents its reducing the soil as quickly as it would, if the harrow was undivided'.[19] Even the single-framed harrow used in county Kilkenny had sometimes to be loaded with stones to increase its weight.[20] This disagreement, like many others, seems to have arisen from a failure to recognise that different types of harrows are required for different tasks. John Wynn Baker's catalogue of 1767 advertised ten types of harrows, but even though light and heavy harrows were distinguished very early, harrow types were probably not sufficiently separated by their two main functions: breaking up fallows (heavy work), and covering seed (light work). This was identified as a problem by at least one writer, in 1804: 'It is ridiculous to suppose, that any one [harrow] . . . can answer all the various purposes, for which it is required, though how seldom do we find the farmer who has an idea, that one of a second form is requisite'.[21]

Draught Animals

From the earliest times horses were the most common draught animals used in harrowing. This is in contrast to ploughing, where as we have seen, oxen were widely used until the sixteenth century. It has been suggested that horses were preferred for harrowing because the draught required was not so heavy as for ploughing, and horses were faster than oxen.[22] The practice of attaching harrows to horses' tails is mentioned in several seventeenth-, eighteenth- and nineteenth-century accounts. However, the advice of some agriculturalists, that oxen should not be used in harrowing, was not always followed. One of the earliest clear illustrations of harrowing shows the implement being pulled by two oxen (Fig. 34).

There are also references in Irish literature of many kinds to men and women pulling harrows. One of the earliest of these comes from the *Annals of Ulster*, where an entry for 1013 A.D. states that Gilla Mochonna, king of south Brega, 'yoked the foreigners under the plough and two foreigners pulling a harrow after them'. In very recent times, by contrast, the story of a 'butter-witch' from near Comber in county Down shows at least a memory of harrows being pulled by people. The witch was shot while trying to escape from a vengeful farmer, but later tried to explain her wounds by claiming that she had fallen while 'trailing a harrow.'[23]

During the later nineteenth and early twentieth centuries a very wide range of harrow types were used in Ireland. The implements manufactured in Irish foundries followed, with little variation, designs developed in England and Scotland. The different functions performed by harrows were reflected in increasingly specialised designs. Disc harrows, zig-zag harrows, spring harrows and drill harrows were widely used on even medium-sized farms by 1900. However, much less specialised implements also remained in use. Harrows with wooden tines, in use in medieval times, were still used in the nineteenth century. As late as 1949, Evans claimed that 'the time is still remembered when the old "six bull" wooden harrow was fitted with wooden tines or cows' horns'.[24] Bush harrows also persisted in use for a very long time. At least one late nineteenth-century text described the function of bush harrows as if these were still in common use. For both spreading manure and covering grass-seed, the light dense coverage of the ground surface by a flattened bush continued to be cheap and efficient, even after web or chain-link harrows were developed in the mid-nineteenth century. As with so many implements and techniques associated with 'common' Irish farming practice, we must be careful not to confuse cheapness and apparent simplicity with crude or inefficient husbandry.

Rollers

Rollers have probably changed less than any other of the horse-drawn implements discussed in this book. It is difficult for museum curators even today to obtain specimens for collections, since many farmers still find horse-drawn rollers efficient, when modified so that they can be pulled behind a tractor. Even these relatively uncomplicated implements provoked some debate amongst agriculturalists, however. Harrows and rollers were often used one after another:

> As harrowing is on some occasions performed before, and on some after sowing, so is rolling: and it is also sometimes used before, sometimes after harrowing; and very frequently and properly between the two harrowings of the same piece of ground.[25]

Rollers complemented the work of harrows, breaking ground into a fine tilth, and compressing seeds into the earth, but they were also used to make the surface smooth, so that care of growing crops, and harvesting, were made easier. They could be used for this last purpose, to a limited extent, even where land was made into ridges. In this case farmers were advised to roll across the ridges.

Throughout the last two centuries rollers could be made from wood, stone or metal. In 1757, a British farming text, republished by the Royal Dublin Society, advised that wooden rollers (Fig. 38) only should be used in fields, as those made of stone or metal would be too heavy, and compressed the soil too much. No diameter was given, but the length recommended was eight feet.[26] Most nineteenth-century writers, however, claimed that wooden rollers were suitable only for light soils, although they were still valued for levelling potato or turnip fields after digging, and also for compressing earth around newly sown grain.[27] Wooden rollers are still sometimes used in western Ireland. Those used around the Bloody Foreland, in north-west Donegal, are constructed from a solid block of hardwood, bound with iron straps at either end to prevent splitting. This construction was recommended in 1893 by Thomas Baldwin. Baldwin criticised the narrow diameter (circa. one foot) of many wooden rollers, however, claiming that this made them difficult to pull and liable to sink into hollows.[28] Some other agriculturalists approved of a narrow diameter, however. Sproule argued that 'by increasing the diameter, the effect is diminished, for the same weight is supported by a greater space'.[29]

Many of the stone rollers still found around the Irish countryside are also narrow in diameter. Some surviving specimens are made of granite, which was the stone said to be most commonly used in the nineteenth century, although sandstone was also used. In 1844, Martin Doyle pronounced granite rollers to be 'very efficacious, as well as extremely cheap, the cost of one being about £1, with a scraper and frame, while the price of a metal roller ranges from £8 to £12'.[30] Stone rollers had the serious defect, however, that they were liable to fracture if the roller was pulled over uneven stone roads or yards.

The rollers most praised by agriculturalists were made from hollow cylinders of cast iron. As with wooden and stone rollers, these could be mounted on a simple frame, or they could be fitted with shafts, either for one or two horses. (Where two horses were used, farmers were advised that they should be harnessed abreast. Horses following in the same track could make the ground where they walked uneven.) One late nineteenth-century text suggested that farmers should buy the iron cylinders from local foundries, and have them mounted on a frame by a blacksmith.[31] Surviving specimens suggest that many small farmers bought rollers which had the dimensions recommended in 1839: that the roller should be five to six feet long, and twenty to thirty inches in diameter.[32]

Metal rollers had important advantages over those made from wood or

Fig. 38. A wooden roller from near Carnlough, Co. Antrim. The weight of this roller could be increased by the farmer sitting on the frame, as shown (WAG 1171).

stone. Their smoothness meant that clods of earth were less likely to cling to them, and even more important, these rollers could easily be made up of two or more cylinders set side by side. The advantage of having rollers made up of several sections was stressed by John Sproule:

> Considerable inconvenience has been experienced in the use of the common roller when turning it at the side of the field; for the cylinder, not moving on its axis, but, being drawn along the surface of the ground, is liable to make depressions, and tear it up before it comes again into the direct line of draught, and at the same time, the labour of turning it round is fatiguing to the animal employed in working it. To remedy this, the cylinder should be divided into two parts, so that each part may revolve separately, and, on turning, one of them can move forward, while the other is taking a retrograde direction.[33]

The best weight for rollers depended on the tasks they were required to do. Weight could be varied by piling stones in a box set on the implement's wooden frame. Farmers also sometimes added their own weight, by standing or sitting on the frame as the roller was pulled (Fig. 38). One late nineteenth-century Irish agricultural writer enthused about a solution to the need for implements of variable weight attempted by Amies and Barford of Peterborough. This firm produced rollers which

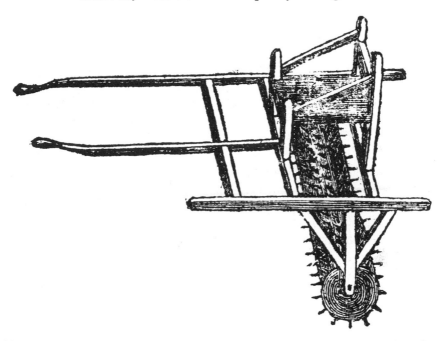

Fig. 39. An early nineteenth-century spiked roller, used on some large Co. Cork farms (*Munster farmer's gazette* (1811), vol. 2, p.137).

could be filled with water, so allowing very fine adjustments to their weight.[34]

The relatively unchanging construction of rollers can be illustrated by the case of the spiked roller. Spiked rollers were probably in use in south-eastern England by 1700,[35] and were being manufactured in Cellbridge, county Kildare, by 1767,[36] although a letter to *The Munster farmer's magazine* in May, 1811, which also included an illustration of a spiked roller (Fig. 39), expressed the opinion that not many of the 'gentlemen' of county Cork had one, 'and it is very certain that no farmers are provided with it.'[37]

The most famous of all spiked rollers, Crosskill's clod crusher, patented in 1841 (Fig. 40), was immediately recognised as very efficient:—

> It was formed by loosely fitting a number of cast iron discs each 30 in. in diameter, along the length of a 5 ft. 6 in. horizontal axle. Each disc was serrated around its entire circumference like a large cog wheel, with additional radial teeth on both of its sides. As the roller was hauled forward by horses, the discs revolved and large clods of earth were broken down by the insertion of the teeth.[38]

Fig. 40. Crosskill's clod crusher, patented in 1841, but still used on large Irish farms around 1900 (Stephens (1871), vol. 2, p.21).

Martin Doyle recommended Crosskill's clod crusher to Irish farmers in 1844.[39] In 1863, the implement was described as 'absolutely necessary on every farm',[40] and in 1893 was still being used on large Irish farms.[41]

Like many other aspects of farm technology, however, even the slow changes in roller design, and the low-key debates between agriculturalists about the best shape and size for the implements were not relevant to many small Irish farmers. In 1812, Wakefield found that throughout much of Ulster the use of the roller was unknown.[42] In 1837, Binns claimed that throughout the barony of Murrisk, county Mayo, there were no rollers used apart from those owned by Lord Sligo.[43] On small holdings, the mells and rakes used instead of harrows also fulfilled the main functions of rollers. Even where horses were used, many farmers broke down soil with 'drags' made from pieces of wood nailed on to a wooden frame. Arthur Young seems to have been describing an implement of this type which was used at Furness, near Dublin. This was described as a 'bull harrow . . . that is, . . . without teeth'.[44]

Sowing

Grain seed could be planted in autumn or spring. Arthur Young thought that, in either case, many of the Irish sowed their crops too late.[45] There were differences in the time of sowing, however, and these continued throughout the next century. In the later nineteenth century, for example, turnips were said to be sown between a fortnight and a month later in the west of Ireland than in the east.[46] During the period covered

Fig. 41. Sowing flax broadcast from a sheet, near Hillsborough, Co. Down, in 1783 (detail of a print by W. Hincks).

in this book, most small farmers sowed their grain and grass seeds broadcast. Seed sown by hand was carried in boxes, baskets, skin trays or sowing sheets. Evans claims that straw seed baskets were used only in south and east Ireland where he believes they were a survival from medieval baskets used in the area known as the Pale.[47] The collection of the Ulster folk and Transport Museum includes seed trays made by stretching a piece of cloth around a kidney-shaped wooden frame. However, simple pieces of sheet, and specially sown cloth bags, seem to have been the most common means for a sower to carry seed (Fig. 41).

In 1839, Sproule gave a clear description of sowing by hand:

> The sheet is kept distended by the left hand, while the right is used in sowing. The sower walks with a measured step along the ridge, scattering

Fig. 42. Sowing grain with a seed-fiddle, Co. Antrim (WAG 3101A).

the seed in a uniform manner across it . . . A skilful and experienced workman regulates the prescribed quantity of seed to the acre with wonderful precision, and distributes the seed over the ground with the most exact equality.[48]

The ability to cast seed in a wide even arc was prized, and a skilled man was often asked to neighbours' houses to sow for them.[49] Some sowers held their sowing hand with the fingers slightly apart. The seed was thrown out through the spaces between the fingers, ensuring that it was more widely spread.[50] Others flung handfuls of seed hard at the ground, so making it bounce and spread out over the ground.

In the early nineteenth century, machines were introduced which assisted farmers to achieve more even broadcast sowing. One form consisted of a wide, narrow seed box with a lid, mounted across a light wheelbarrow frame. As the machine was pushed, a long axle fitted with brushes or iron teeth revolved inside the seed box and pushed seed out through evenly spaced holes. This basic form persisted over a long period and was used especially for sowing grass-seed. The length of the seed-box was increased until the 'convenient and manageable' length of sixteen feet was arrived at.[51] A horse-drawn version of these broadcast sowers was recommended to Irish farmers by John Sproule in 1839,[52] but

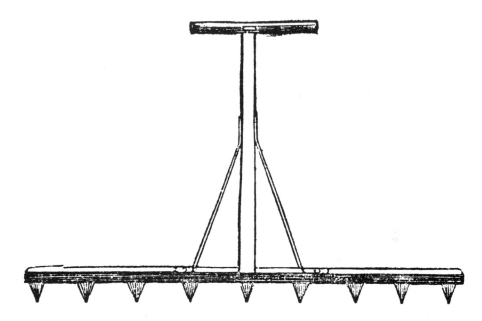

Fig. 43. A dibbler with nine wooden points (*Farmer's gazette* (1844), vol. 3, p.881).

the implements do not ever seem to have come into common use in Ireland.

One broadcast sowing machine which did become very popular, however, was the seed fiddle (Fig. 42). The early origins of fiddles are obscure. It has been suggested that they were developed in England some time during the eighteenth century,[53] but they are not described in either Irish or British standard farming texts before the early twentieth century.[54]

Fiddles are still used in Ireland, especially for sowing grass. Ensuring an even spread of seed is relatively easy. The sower takes regular strides, and pulls or pushes the bow in time with each step. *Stephen's book of the farm* estimated that it was possible to sow up to four acres of grain an hour using a fiddle, while three could be sown easily.[55]

Experiments aimed at developing techniques of sowing where seed could be planted in evenly spaced rows have been carried out for at least three and a half centuries. The practice of dibbing or dibbling seed, which involved making individual holes into which individual seeds could be set, seems to have been first proposed in England by Edward Maxey in 1601, and became fairly common in the southern counties during the eighteenth and nineteenth centuries.[56] In the 1770s, Arthur

Fig. 44. A horse-drawn seed drill manufactured by Ritchie of Ardee, Co. Louth (*Farmer's gazette* (1847), vol. 6, p.473).

Young recommended that beans should be sown using dibbers:

> Women take some beans in their aprons, and with a dibber pointed with iron make the holes along the string [marking the line to be sown] with their right hand, and put the bean in with their left . . . The beans are put three inches asunder, and two or three inches deep.[57]

Dibblers with rows of wooden points fixed to the bottom of a wooden frame (Fig. 43) remained in use in vegetable gardening well into the present century.[58] The use of dibblers known as steeveens (Irish: *stibhíní*) for planting potatoes will be discussed in the chapter on that crop. However, the main significance of dibbing was that by planting in rows, hand-hoeing amongst young plants was much easier, so that crops could be kept weed-free, and soil aerated. Dibbing therefore pointed the way to the major development in sowing seed during the last two hundred and fifty years: the use of machines to sow seeds in rows or drills.

The most famous inventor of seed drills, perhaps indeed the most famous name in the eighteenth-century English 'Agricultural Revolution', was Jethro Tull. In fact, Tull was not the first to invent a grain drilling machine, but he certainly devised a practical one. Most

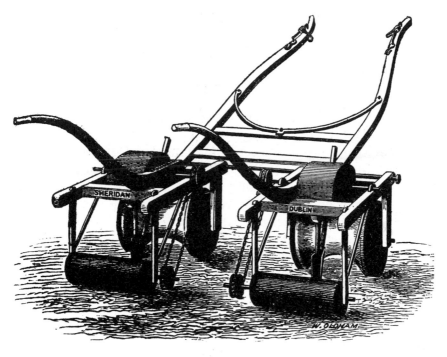

Fig. 45. A turnip-sowing machine, manufactured by Sheridan of Dublin (*Farmer's gazette* (1848), vol. 7, p.187).

horse-drawn grain sowing machines consisted of a large seed box fixed above the machine's wheels, and a series of tubes through which seeds fell at fixed intervals, so that they were sown in long straight, equidistant rows (Fig. 44). This basic construction remained common, and by the 1860s seed drills were in all essentials the same as those used in modern times.[59] Small hand-operated seed drills for sowing grain were also developed, mainly for filling in spaces in drills where germination had not been successful, and drills were also made for sowing beans and other vegetables. From the evidence of surviving specimens, turnip sowing machines seem to have been fairly commonly used in Ireland in the later nineteenth century. John Sproule described the construction of a turnip sowing machine, in 1839 (Fig. 45):

The machine drawn by a horse sows two drills at each turn, . . . the seeds are put into cylindrical boxes of iron or tin, which are made to revolve during the time the machine is in motion, and being perforated in a line all round, the seeds fall through as the boxes revolve. Funnels are placed below each of these boxes, and through these the seeds are carried to the ground in tubes defended at their lower part by coulters, which make ruts in the ground into which the seeds fall. [The machine has] two light wooden rollers which follow each track of the coulters and cover the seeds . . . [and also a large wooden roller in front]. The effect of this roller is to flatten and compress these drills just before the seeds are sown.[60]

The spread of grain drilling machines seems to have been very slow throughout the eighteenth century. By the early nineteenth century, some machines were being manufactured in Ireland. A letter to the Cork Institute in 1813 records the purchase of a 'five row sowing and horse hoeing drill' from the Summerhill Manufactory, Dublin.[61] In 1839, however, drilling was said to have made little progress in Ireland, and this claim was repeated in 1854.[62] By the 1860s, machines for both broadcast sowing and drilling were said to be 'rapidly superceding sowing by hand'.[63] Even after this date, however, horse-drawn seed drills seem to have been confined to larger farms in established tillage areas.

There was widespread debate during the eighteenth and nineteenth centuries on the relative merits of broadcasting and drilling seed. In 1814, an article in the *The Munster farmer's magazine* listed the advantages of drilling grain seed:

1. Seed was saved.
2. Labour was saved. (If seed was ploughed in, as described in the last chapter, a ploughman and a pair of horses could cover an acre a day. A seed drill with an operator and one horse could sow three acres.)
3. The increase of speed possible using a drilling machine meant that sowing could be carried out as soon as weather improved at all.
4. Crops resulting from drilling were claimed to be at least equally productive to those grown from broadcast seed.
5. Drilled grain ripened more evenly, because the seed had been at an equal depth.
6. Drilled crops could be weeded much more easily than those sown broadcast.

The 'member of the Barrymore farming society' who wrote the article in which these advantages were listed had discovered them after using a machine made by J. Barry of Glanmore, whom he recommended to his fellow farmers.[64] In 1839, however, Sproule argued that while for the cultivation of green crops drilling was one of the most important improvements of modern agriculture, the value of drilling grain crops might have been exaggerated. He believed that drilling was advantageous in light soils but argued that, in the end, the value of a grain crop was determined by how much it tillered (or multiplied the number of its stalks) during growth. Drilling seed did little to encourage this proliferation.[65]

In 1844, Martin Doyle wrote that 'drill husbandry is the most perfect mode of culture, and we hope to see it widely extended in Ireland'.[66] Even Doyle, however, allowed exceptions to this. Barley, he believed, did not require hoeing, which drill cultivation made easier, and probably tillered much better if the seeds were evenly spread over the surface by broadcasting, than if they were sown in drills where they were 'thrown so much together that they cannot tiller so freely'.[67] Another instance in which Doyle thought broadcast sowing might be best was when wheat was being planted in autumn, following a potato crop:

> In very stiff soils, and at the season of removing potatoes, the weather is often so moist, and the soil so adhesive as to render drilling difficult, if not impracticable; and then the most expeditious broadcast method must be resorted to. By this the seed is either ploughed under, in ridges on land in the fallow state (as is generally the case in the best wheat counties in Ireland) or harrowed in, after the ploughing, as is usual in England. The advantage in our judgment is greatly in favour of the Irish mode, which gives a far better and deeper covering to the seed, and thus prevents the plants from being thrown out in the spring, as frequently happens after severe frost.[68]

Doyle's discussion of this last case brings us back to points which will be emphasised throughout this book. In the end, techniques of sowing seed cannot be isolated from other cultivation techniques. The way in which ground was turned had direct relevance for the method of planting. High-crested work meant that seed tended to fall between furrow slices, and so grew up in rows. The same effect could be produced by the technique known as 'ribbing'. This was recorded by Tighe in county Kilkenny in 1802,[69] and was described by Thomas Baldwin in 1874 as a favourite mode of sowing wheat: 'The ground is made into ridglets or ribs, with a common plough stripped of its mouldboard, or with a ribbing machine. On cottier farms the ribs may be formed with the corner of a hoe or with a small shovel. The seed is scattered broadcast on the ribbed surface, and covered with a harrow'.[70]

'Common' techniques of sowing grain cannot simply be opposed to the drilling of seed as inefficient to efficient, or simple to complex. The spread of seed drilling machines was very slow even among 'improving' farmers. The reasons for this can be found in calculations of farmers, large and small, who took into account not only labour and draught requirements, and local environmental conditions, but also the ease with which new techniques could be fitted into their existing technical systems.

NOTES

1. Bell, J., 'Harrows used in Ireland', *Tools and tillage,* vol. 4:4 (Copenhagen, 1983), p.195.

2. Bincy, D.A. (ed.), *Críth Gablach* (Dublin: Stationery Office, 1941), p.16.

3. O'Loan, John 'The Manor of Cloncurry, Co. Kildare, and the feudal system of land tenure in Ireland', Department of Agriculture (Ireland), *Journal,* vol. 58 (Dublin, 1961), pp.15-19.

4. Doyle, Martin *Hints to smallholders on planting and on cattle, etc.* (Dublin, 1830), p.71.

5. Thompson, R., *Statistical survey of the county of Meath* (Dublin, 1802), p.116.

6. Tighe, W., *Statistical observations relative to the county of Kilkenny* (Dublin, 1802), p.300.

7. McEvoy, John, *Statistical survey of the county of Tyrone* (Dublin, 1802), p.47.

8. Bell, J., *op. cit.,* p.197.

9. Baker, John Wynn, *A short description and list, with the prices of the instruments of husbandry, made in the factory at Laughlinstown, near Cellbridge, in the county of Kildare* (Dublin, 1767-1769), p.30.

10. Hale, Thomas and others, *A compleat body of husbandry,* vol. 2 (Dublin, 1757), p.105.

11. Tighe, W. *op. cit.,* p.500.

12. Dutton, Hely, *Statistical survey of the county of Clare* (Dublin, 1808). p.63.

13. Hale, Thomas, *op. cit.,* p.104.

14. Doyle, Martin, *op.cit.,* p.71.

15. Bell, J., *op. cit.,* p.200.

16. Baker, John Wynn, *op. cit.,* p.12.

17. Dutton, Hely, *A statistical and agricultural survey of the county of Galway* (Dublin, 1824), p.90.

18. Bell, J., *op. cit.,* p.200.

19. Thompson, R., *op. cit.,* pp.116-117.

20. Tighe, W., *op. cit.,* p.301.

21. Coote, C., *Statistical survey of the county of Armagh* (Dublin, 1804), p.169.

22. O'Loan, John, 'A history of early Irish farming', part 2, Department of Agriculture (Ireland), *Journal,* vol. 61 (Dublin, 1964), p.19.

23. Bell, J., *op. cit.,* p.202.

24. Evans, E. E., *Irish heritage* (Dundalk, 1949), p.91.

25. Hale, Thomas, *op. cit.,* p.112.

26. *Ibid.,* p.114.

27. Doyle, Martin, *A cyclopaedia of practical husbandry,* rev. ed. W Rham (London, 1844), p.497.

28. Baldwin, Thomas, *Introduction to practical farming,* 23rd ed. (Dublin, 1893), pp.155-156.

29. Sproule, John, *A treatise on agriculture* (Dublin, 1839), p.70.

30. Doyle, Martin, *op.cit.,* (1844), p.497.

31. Baldwin, Thomas, *op. cit.*, p.157.

32. Sproule, John, *op. cit.*, p.701.

33. *Ibid.*, p.71.

34. Baldwin, Thomas, *op. cit.*, p.157.

35. Fussell, G.E., *The farmer's tools* (London, 1952), p.68.

36. Baker, John Wynn, *op. cit.*, p.29.

37. *The Munster farmer's magazine*, vol. 2 (Cork, 1813), pp.137-141.

38. Partridge, Michael, *Farm tools through the ages* (Reading, 1973), p.85.

39. Doyle, Martin, *op.cit.*, (1844), p.497.

40. Purdon, *op. cit.*, p.308.

41. Baldwin, Thomas, *op. cit.*, p.156.

42. Wakefield, E., *An account of Ireland, statistical and political*, vol. 1 (London, 1812), p.362.

43. Binns, J., *The miseries and beauties of Ireland*, vol. 1 (London, 1837), pp. 391-392.

44. Young, Arthur, *A tour in Ireland*, vol. 2 (Dublin, 1780), p.208.

45. *Ibid.*, vol. 1, p.72.

46. Baldwin, Thomas, *op. cit.*, p.15.

47. Evans, E.E., *Irish folkways* (London, 1957 (1967)), pp.142-143.

48. Sproule, John, *op. cit.*, pp.245-246.

49. Interview with Cormac McFadden, Roshin, Co. Donegal (Ulster Folk and Transport Museum Tape R80.39).

50. Interview with John Joe McIlroy, Derrylea, Co. Monaghan (Ulster Folk and Transport Museum Tape R85.143).

51. Partridge, Michael, *op. cit.*, p.113.

52. Sproule, John, *op. cit.*, p.246.

53. Partridge, Michael, *op. cit.*, p.119.

54. *Stephens' book of the farm*, vol. 2, 5th ed. rev. J. Macdonald (Edinburgh, 1908), p.124.

55. *Ibid.*

56. Partridge, Michael, *op. cit.*, p.120.

57. Young, Arthur, *op. cit.*, vol. 2, p.120.

58. Evans, E.E., *op. cit.* (1957 (1967)), p.146.

59. Fussell, G.E., *op. cit.*, p.113.

60. Sproule, John, *op. cit.*, pp. 71-73.

61. *The Munster farmer's magazine*, vol. 3 (Cork, 1814), p.35.

62. Sproule, John, *op. cit.*, p.247; Sproule, John (Ed.), *The Irish Industrial Exhibition of 1853; a detailed catalogue of its contents* (Dublin, 1854), p.211.

63. Purdon, *op. cit.*, p.240.

64. *The Munster farmer's magazine*, vol. 3 (Cork, 1814), p.133.

65. Sproule, John, *op. cit.* (1839), pp.246-247.

66. Doyle, Martin, *op. cit.* (1844), p.215.

67. *Ibid.*, p.45.

68. *Ibid.*, p.569.

69. Tighe, W., *op. cit.*, pp.200-201.

70. Baldwin, Thomas, *Introduction to Irish Farming* (London, 1874), p.26.

CHAPTER 6

Potatoes

The cultivation of potatoes has been by far the best-known, or notorious, aspect of Irish farming. Historians and economists have argued at great length about the significance of the crop. It has been pointed out that there was never a time when the Irish ate potatoes and nothing else. Variations of diet with social class, and also between different regions, have been emphasised. However, despite these reservations, historians are generally agreed that the failure of the potato crop in the years 1845 and 1846 triggered the greatest catastrophe in recent Irish history, and that the economic, political and social results of the tragedy are still having deep long-term effects today[1] (Fig. 46). Over one million people died of hunger and disease in the Great Famine, and many more emigrated in the decades following it. This decimation has left Ireland in the unique position amongst European countries of having a population less than half of its mid-nineteenth century size.

It has been convincingly argued that the famine and its aftermath can only be understood by examining the wider political and economic structures of nineteenth-century Ireland. Historians have aimed at showing how the rapid increase in Ireland's population before the mid-nineteenth century, the various elements in the systems of landholding, and the development of industry or a lack of it, all made the position of the Irish poor so vulnerable that the failure of a single crop could have such an effect.[2] While recognising the importance of these debates, however, this chapter will concentrate on technical aspects of growing potatoes. In the end, these have implications for the big historical questions, but in this book we can only suggest possible connections, rather than trace them in detail.

Potatoes were probably introduced to Ireland from America in the late sixteenth century. Despite a tradition that Sir Walter Raleigh cultivated the plants in his garden at Youghal, county Cork, it seems equally likely that the first tubers may have come via Spain.[3] One of the earliest definite references to the crop dates from 1606, when Sir Hugh Montgomery's wife provided Scottish workers with land near Comber, county Down, on which to plant potatoes.[4] The crop seems to have

Fig. 46. Carrying away the dead in Co. Cork, 1847.

become a central part of the Irish diet very quickly. In 1689, John Stevens wrote that, 'None but the best sort . . . eat wheat or bread baked in an oven or ground in a mill; the meaner people content themselves with little bread, but instead thereof eat potatoes which with sour milk is the chief article of food'.[5]

Potatoes contain carbohydrates, protein, and mineral salts. Combined with milk, they could provide a liveable diet. Arthur Young recognised this in 1780, and also that apart from nutrition, potatoes had the advantage that large amounts could be produced from a small area. 'They have all a bellyful of food whatever it is, as they told me themselves; and their children eat potatoes all day long, even those of a year old'.[6] 'One acre does rather more than support eight persons the year through, which is five persons to the English acre. To feed on wheat, those eight persons would require . . . two Irish acres, which at present imply two more for fallow, or four in all'.[7] It has been estimated that in modern Ireland an acre can yield up to nine tons of potatoes, enough to feed six adult men for a year. Before the famine, consumption may have reached eight pounds a day per person.[8]

Potatoes were used not only for human consumption, but also as animal fodder, especially for pigs. Arthur Young viewed the eating habits of the Irish poor with some derision, but also recognised their virtues:

> Mark the Irishman's potato bowl placed on the floor, the whole family upon their hams around it, devouring a quantity almost incredible, the beggar seating himself to it with a welcome, the pig taking his share as readily as the wife, the cock, hens, turkeys, geese, the cur, the cat, and perhaps the cow — all partaking of the same dish. No man can often have been a witness of it without being convinced of the plenty, and I will add, the cheerfulness that attends it.[9]

A close similarity between animal and human diet persisted in some cases even into the present century. The Donegal writer, Patrick MacGill, described with bitterness the experience of a young hired farm servant in county Tyrone:

> In the morning I was called at five o'clock and sent out to wash potatoes in a pond near the house. Afterwards they were boiled in a pot over the kitchen fire, and when cooked they were eaten by the pigs and me. I must say I was allowed to pick the best potatoes for myself, and I got a bowl of buttermilk to wash them down. The pigs got buttermilk also. That was my breakfast during the six months. For dinner I had potatoes and buttermilk, for supper buttermilk and potatoes.[10]

In the early nineteenth century some observers believed that potatoes could provide a solution to the problem of providing food for Ireland's growing population. There was a continuing debate as to whether farmers, or landlords, should be allowed to sub-divide farm holdings. In the second chapter of this book, we have seen also that from the mid-eighteenth century there was a large-scale movement of poor tenant farmers and cottars on to marginal land. This movement was assisted by the relative ease with which potatoes could be grown:

> Of the different modes adopted in the bringing in of soils hitherto uncultivated, the spade potato culture is the most encouraging, which in the course of a few months, not only gives a return on food, but leaves the land in a state of producing more . . . To the potatoe . . . [these green mountains] ought to be devoted; if grain will ever prosper on them it must be under the auspices of that vegetable; if it does not, still it ought to be resorted to as a preparation for grass.[11]

Just before 1845, there were about two million acres of potatoes cultivated in Ireland, mostly on plots of less than one acre.[12]

A great variety of potatoes have been cultivated during the last two centuries. Some of the most popular are listed below:

The Irish Apple. This variety was certainly grown by 1770. It was popular for its dry mealy consistency when cooked, and also because it kept well in storage, so providing food for the 'hungry months' of July and August, just before the next year's crop had matured. However, it does not seem to have been grown much after 1846.

The Yam. This seems to have been a very popular variety amongst poorer farmers, as early as 1808. However, it was very susceptible to blight, and may have added to the scale of the destruction of the potato crop in the 1840s.

The White Kidney. Another popular variety in the early nineteenth century, particularly because it could be harvested early, and so provided food during July and August.

The Cups. This was a luxury variety, grown as early as 1808. It was thought to be particularly nutritive, but was criticised because it 'stay[ed] too long in the stomach'.

The Champion. This variety was developed in 1863, and introduced to Ireland in 1876. Until 1894, almost 80% of the potatoes grown in Ireland were Champions, and they survived in use well into the present century. A resistance to blight was one of the Champion's main advantages, but in 1890 they too succumbed to the disease. [13]

Planting Potatoes

It is remarkably easy to get a potato plant to grow (one county Tyrone farmer remembers seeing the shoot of a young plant which had forced its way through the leather sole of a shoe thrown on a midden). [14] Obtaining high yields, however, is much more difficult. Discussions aimed at improving methods of potato cultivation during the past two centuries have covered every stage between planting and harvest.

One way to obtain a new crop of potatoes was simply to leave some small tubers produced by the previous year's plants in the ground, so that they grew the following year. Arthur Young disapproved of this practice which he recorded near Annsgrove, county Cork: 'Potatoes they plant in a most slovenly manner, leaving the small ones in the ground of the first crop, in order to be seed for the second, by which means they are not sliced: sometimes a sharp frost catches them, and destroys all these roots'. [15] At Castle Lloyd, county Limerick, he also found that 'they . . . [trust] to the little potatoes left in the ground, and which they spread in digging; but this is a most slovenly practice; if they were to plant the second crop it would be better than the first'.[16]

Planting in a more positive way could begin as early as February, and continue until May. Potato seed or 'sets' could consist of whole, or sliced up, tubers. There is also at least one recorded instance, where only the young shoots were used:

> I met in this town [Limerick] a certain reverend doctor, inventor of a new method of growing potatoes. This consists in cutting out, in spring, the shoots or eyes and planting them. It appears that the result is just as good as if the potatoes were cut up and planted, and with this benefit, that the tubers furnishing the shoots are still available as food — for pigs at any rate.[17]

Early nineteenth-century writers, however, generally accepted that large sets, or whole potatoes, led to increased yields.

Ridges, constructed using spades or ploughs, or both, were widely used for potato cultivation. Potatoes could be planted while ridges were being made. On lazy-beds, the tubers could be set on the untilled strips of grass which formed the base of these ridges, and sods then turned over on top of them. On the other hand, Arthur Young recommended the practice of setting potatoes in the furrows formed by lea ploughing, so that the next furrow slice turned covered the seed:

> In [the] . . . first ploughing, which should begin the latter end of February, or the beginning of March, the potatoes are to be planted. Women are to lay the sets in every other furrow, at the distance of twelve inches from set to set close to the unploughed land, in order that the horses may tread the less on them. There should be women enough to plant one furrow in the time the ploughman is turning another.[18]

The spacing of potato seed varied widely. In 1807, Rawson suggested that potatoes should be set so that each plant would be surrounded by a square yard of soil.[19] In 1814, in Dunaghy parish, county Antrim, however, sets were placed only six to eight inches apart, although farmers wanting to grow larger potatoes placed the seed from ten to twelve inches apart,[20] while at Athy, county Kildare, Arthur Young found that sets were placed fifteen inches apart.[21] In this latter case the seed was planted using a dibbler. Potato dibblers varied in size from small hand-held sticks, to larger implements which were provided with a foot-rest. The best-known of these was the steeveen (Irish: *stibhín*).[22] Steeveens were made of wood and had a foot-rest either cut into the shaft, or made from a separate piece of wood (Fig. 47). They were used

Fig. 47. Mr. Ronnie Cromie demonstrating the use of a steeveen, in the grounds of the Ulster Folk and Transport Museum (UFTM, L1501).

mostly in north Connaught and south-west Ulster. Even within these regions, however, dibblers were by no means universally employed. As mentioned earlier, potatoes could simply be set on the ground, and covered either with a turned sod, or by loose soil.[23] Farmers also sometimes used their spades to make holes in the ridge into which the seed was placed. This technique was known as kibbing. In 1814, Sampson recorded the practice in county Derry:

> In the operation of kibbing, the labourer is provided with an open bag of potatoes ready cut for seed, which he binds round his waist so as to have it ready to take out the seeds one by one. These he throws with great exactness into an hole, which he first makes by entering his spade forward into the soil, in a sloping direction; in raising the handle of the spade toward a perpendicular, a vacant space in the soil is left open, behind the spade-iron; into this vacuity the seed-potatoe is thrown . . .; the spade being instantly withdrawn, the earth falls back to its level, and the potatoe plant is at the same time covered. An expert kibber will cover twenty seeds in one minute.[24]

One man, who has planted potatoes in county Fermanagh by kibbing, said that it was particularly useful if the tubers had well-grown shoots, since it ensured that these would be well covered, and so protected against frost damage.[25]

Just as potato seed could be planted at different stages in ridge construction, so manure could be added before, during, or after the ridge had been made. At Charleville, for example, Arthur Young found that the poor would 'sometimes put a little dung or [limestone] gravel on the grass, and plant it with potatoes'.[26] In county Cork in 1813, on the other hand, it was claimed that 'the lower class' would sometimes dibble their potato seed into ridges, and spread the dung as a final stage.[27] At whatever stage manure was added, however, contemporaries agreed that very large quantities were used. Arthur Young frequently noted that the poor used all their manure for their potato patches.[28] Farmers sometimes rented land to cotters already dunged,[29] or let them have it free on condition that they manured it for potatoes.[30] The heavy use of farmyard manure persisted into the present century. An agricultural inspector reported that in county Carlow, for example, 'the potatoe has the claim upon the manure heap and invariably gets a fair dressing of the best dung available'.[31] At the same time another inspector criticised farmers in south Wexford for relying too much on farmyard manure, 'heavy dressings of which give unsound crops'.[32] The quantities of manure required when potatoes were grown in beds or ridges was a central criticism put forward by advocates of the drill cultivation of potatoes.

Drill Culture

Potatoes were cultivated in what were known as raised drills. The ground was prepared, often by deep ploughing, cross ploughing, and harrowing. Raised drills could be made using spades, lea ploughs and shovels, but 'drill' ploughs were also developed (Fig. 48). The main characteristics of these were that they had long narrow shares or 'points' and two straight mouldboards, and that they had no coulters. When pulled through the ground, the mouldboards pushed soil up in heaps on either side (Fig. 49).

An illustration taken from a nineteenth-century British text (Fig. 50) clearly shows the different stages in planting potatoes in drills. In 1893, Thomas Baldwin summarised these operations: 'In the hollows of the drilled ground, manure is spread; the "seed" or "sets" are placed over it at intervals of ten to twelve inches, according to the richness of the

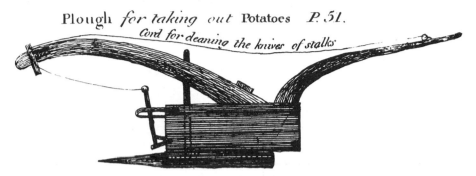

Fig. 48. A wooden drill plough from Co. Down (Dubordieu, J. (1802), p.52).

Fig. 49. Mr. Bertie Hanna of Saintfield, Co. Down, covering potatoes with a metal drill plough, in the grounds of the Ulster Folk and Transport Museum (UFTM, L2340/2/12).

Potato-planting.

Fig. 50. Stages in planting potatoes in drills, as illustrated in a classic Scottish farming text: a. Ploughman making drills. h,k,n,o,p. Spreading manure from a cart. r,s. Planters. u. Ploughman covering potatoes with a drill plough (Stephens (1871), vol. 1, p.513).

ground and the variety of potatoes grown; and both the manure and seed are covered by splitting the drills'.[33] In the early nineteenth century it seems to have been common to level the field with harrows after sowing. One reason given for this was that it made the plants grow straight up out of the ground, whereas if the earth was left heaped above them in drills, they would tend to come through the sides.[34]

After planting, the growing potato plants were weeded. Early writers complained that this was not done enough, but some weeding by hand, and hand-hoeing of beds and ridges were recorded.[35] Several horse-drawn implements were developed in the late eighteenth century for weeding drills. These included harrows, horse-hoes, and grubbers (Fig. 51). Most were constructed around a triangular frame whose width could be adjusted so that the implement could be used on drills of different dimensions, without damaging the growing plants. The plants were also 'moulded up', sometimes several times. This involved heaping soil up around the stems to encourage the growth of the tubers. Moulding up usually followed weeding and grubbing, which loosened the soil between

Fig. 51. Mr. Bertie Hanna grubbing potato plants in the grounds of the Ulster Folk and Transport Museum (UFTM, L1608/4).

drills. Shovels were often used for the task, but the double mouldboard drill plough was the standard implement when horse power was utilised (Fig. 52).

John Wynn Baker's factory at Cellbridge, county Kildare, was producing implements for drill husbandry in the 1760s.[36] At Lesly Hill, county Antrim, Arthur Young found that a Mr. Hill had tried Baker's implements, but the experiment had not been a success:

> Mr Hill practised the drill husbandry several years, in consequence of the recommendations of Mr. Wynn Baker. He bought of him a complete set of tools for the purpose, a drill plough, horse hoes, etc., and spared neither attention nor expense to give it a fair trial, but found that it would not answer at all, and then gave it up.[37]

At Derry, county Tipperary, however, Young found that potatoes were being grown in drills, apparently successfully, but without the standard equipment manufactured at Cellbridge. Stubble ground was first ploughed two or three times, and ploughs were then used to make the drill trenches. Dung and potato sets were laid down, and covered using either ploughs, or shovels. The drills were kept clean by constant moulding up with ploughs and shovels.[38]

E

Fig. 52. Mr. W. J. McCann of Clandeboye, Co. Down, moulding up potatoes in the grounds of the Ulster Folk and Transport Museum (UFTM, L1607/8).

Agriculturalists, while admitting that lazy-beds had value on newly broken ground, were adamant that potato drills were in general far superior. Indeed, John Wynn Baker had to defend himself against the charge that the drill husbandry of grain and root crops was all that he was interested in:

> I have lately discovered, that some Gentlemen have formed an idea, that my undertakings have been principally calculated to establish the drill husbandry in Ireland. I shall beg leave to remove that idea, by reminding them, that the introduction of the drill husbandry is only one branch of the general plan.[39]

By the early nineteenth century, however, farming societies were encouraging the cultivation of potatoes in drills without apology. Many of the arguments used in support of the practice echo those put forward by advocates of the drill cultivation of grain crops, which were summarised earlier. Potato drills were claimed to save labour, time, seed and manure. The crops were argued to be easily tended while growing, and more easily harvested. Tighe's dismissal of objections to drilling potatoes in county Kilkenny shows the determination with which the practice was promoted:

> The drill planting has made a slow progress in this district, though a

general practice in Wexford; a few gentlemen practise it, and some farmers, particularly in Galmoy, who learned it from the Palatines in Tipperary; some who have tried it, state ill-founded objections; they say saving manure is disadvantageous to the succeeding crop; but the quantity of manure saved might be added afterwards, with great effect, to the succeeding crop, . . . they say the produce is not as great as in the old way, and the potatoes are too near the surface, and liable to be injured; these objections only spring from their defective mode of execution, by making the drills too close, by not ploughing straight, and by not earthing sufficiently.[40]

Farming societies rewarded good examples of drilling. In 1810, for example, the Cork Institution offered a premium to 'The two farmers who shall cultivate in drills in the best manner, the greatest number of acres of potatoes'.[41] Small-scale projects were also encouraged. In 1814, the Institution gave a premium to a cottager, 'Elinor Leary of Ballyorban, for drill potatoes cultivated by herself and a boy, her son, with a spade and shovel, the drills were marked with a line'.[42]

Members of the Cork Institution also questioned seventeen men who had won premiums for the drill cultivation of potatoes, on the techniques they had used, and the merits of the system in general. Five of these men worked as ploughmen for 'gentlemen' and the other twelve were 'working farmers'. It was found that while several of the men had used double mould board ploughs to make the drills and cover the seed, others had used Scotch or Irish ploughs. The soil needed for moulding up had been loosened in a variety of ways by hoes, bush harrows, and in one case, using the local type of mattock, graffans (Irish: *graffáin*).[43]

One of the men questioned made a statement with which all of the others agreed: 'I have no doubt of the superiority of culture in drills over that in beds. The produce of the former is at least equally great and the quality of the potatoes better'.[44] All agreed that drills left the ground in better condition than beds did, although in one case only one third of the manure was put in the drills that had previously been put in beds. Another witness calculated that drills saved between one third and one half of the labour, one third of the dung, and one third of the seed which would have been used for cultivation in beds. The only instances where beds were thought better were in newly cultivated ground, or in very wet ground.[45]

It is difficult, as with any change in agricultural practice, to discover the rate at which the drill cultivation of potatoes replaced the use of lazy-beds, but impressionistic evidence suggests that there was a large-scale adoption of the new methods during the early nineteenth century. In 1834, for example, it was claimed that, 'The drill has so completely

Fig. 53. Mr. Patrick Doran, and his daughter, making potato drills using a wooden Mourne drill plough, near Attical, Co. Down, in 1962 (UFTM, L1864/5).

superceded the lazy-bed that in the line extending from Dublin by Athlone to Galway, and thence by Ennis and Limerick to Dublin again, we saw but very few fields cultivated in the old way'.[46]

Nineteenth-century Irish foundries, and local blacksmiths, produced large numbers of metal drill ploughs, grubbers and horse-hoes, all for drilling potatoes. Wooden drill ploughs were also made, and in a very few areas, such as around the Bloody Foreland (*Cnoc Fola*) in north-west Donegal, some of these are still in use. Around the south-eastern slopes of the Mourne mountains in county Down, an evolution of drill ploughs occurred which may have been unique to that area[47] (Fig. 53).

'Mourne' drill ploughs seem to have been developed from ploughs similar to the 'Moira' lea plough, which was described in Chapter 4. Modifications to the structure of these ploughs probably began when the techniques of drill husbandry became known in the area. Rather than buy, or make, a double-boarded drill plough, local farmers adapted their lea ploughs for making drills by removing the coulter and lashing a piece of wood on to the plough's land side. This block acted as a second mouldboard, pushing loose earth up into drills. During the nineteenth century, the block, known as the 'false' or 'wee' rest, became more permanently attached to the plough, often lightly nailed in position and only removed when furrow slices were being turned. When metal swing

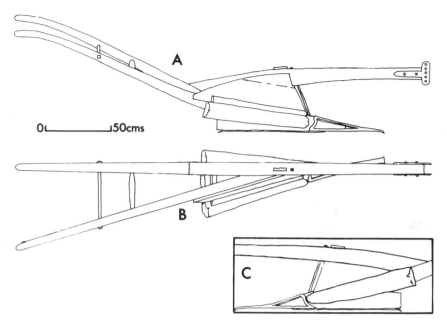

Fig. 54. An asymmetrical drill plough from the Mourne area, Co. Down. A. Side, showing mouldboard, and 'big reest'. B. Plan. C. Land side, showing 'wee reest' (UFTM, L2331/4).

ploughs became common in Mourne in the second half of the century, the ground-turning function of the local ploughs became obsolete. Coulters were no longer attached to the ploughs and false reests became permanent fixtures (Fig. 54). This created an asymmetrical type of drill plough which continued to be made by carpenters and blacksmiths in Mourne until the 1950s.

A farmer using a Mourne drill plough held the left handle slightly lower than the right to counteract the asymmetry of the plough's body. The tendency of the plough's broad body and roughly finished reests to collect soil was used to advantage. The collected soil was shaken off at the end of each drill, and was later used for moulding up. If the plough hit a stone, the farmer shook off some of the soil, and so covered any seed lying near the stone. Despite its odd construction, several farmers who used Mourne ploughs until recently said that they preferred them to metal drill ploughs, especially for making drills in dry, stony soil.

Despite the large-scale adoption of drill culture, however, the use of lazy-beds has also persisted in Ireland, especially on small farms situated in areas of marginal land. On many of these holdings family labour was more than sufficient to perform the work, and the saving in time claimed for drill cultivation was also largely irrelevant, since as we have seen in other chapters, ridges could be made by stages during the course of several weeks, or even months. Manure could be a problem if the family did not have a cow, but if dung was available from even one animal, there was sufficient to manure a crop. By using hand-labour, the farmer was spared the expense of keeping a horse, and could ensure yields which were at least as great as those produced by larger-scale techniques.

Harvesting Potatoes

In Irish folklore, the date most quoted for the beginning of the potato harvest was Lammas day (Irish: *Lá Lúnasa*). This could be the last Sunday in July, or the first Sunday in August.[48] It was sometimes thought to be unlucky to dig the crop before this date, or to reflect bad management by the farming family. Digging the crop could be delayed much longer than this, however. At Kildare, Arthur Young found that cottagers were obliged to clear their potato land by the first of November, after which it was ploughed by the farmer and sown with winter wheat.[49] At Annsgrove, county Cork, however, where there was no such pressure, potatoes were left in the ground until nearly Christmas.[50] while at Nedeen he found that 'the climate in these parts of Kerry is so mild, that potatoes are left by the poor people in the ground the whole winter through'.[51]

Potatoes cultivated in beds could be dug out using spades or forks. One such fork (Fig. 55) was advertised in the *Dublin penny journal* in 1834. The maker, 'J.M.' of 27 Frederick Street, north Dublin, described the implement:

> The flattened portion of each prong is about five inches in length, by one in breadth, thinner at each edge than in the middle, and with spaces of one inch between each prong. They are made of scrap or Swedish iron, (occasionally I make them entirely of steel, in which case they are very light and handy, and wear much longer than when made of iron). They are made with a prong to go into the handle of about four feet in length, and are found much more convenient for digging potatoes, and also for pointing borders in a garden, than any spade.[52]

In large fields, ridges or drills were often opened using a plough. A Mr. Kenney described the arrangement of his workers in a field in county Cork in 1810. The potatoes had been cultivated in beds four feet wide, separated by trenches two feet wide:

Fig. 55. A three-pronged potato digging fork (*Dublin penny journal* (1834), vol. 3, p.283).

The ends of the beds, inconveniently situated for ploughing out, are previously dug with spades, and the stalks cleared away from the beds. Twelve labourers with spades are provided, each attended by an active picker. The bed to be cleared is measured into six equal portions, one of which is allotted to each pair of labourers. A common plough, so set as to penetrate to the deepest potatoe, makes a cut along one side of the bed. The pickers of the first pair of labourers follow the plough quickly, and having collected such potatoes as have appeared, return to the labourers, who are by this time at work back to back, each beginning at an extremity of the portion of the bed allotted to them. As the plough proceeds, the other labourers and pickers to work in the same manner. The plough returning cuts off a slice at the opposite side, which undergoes a similar course of treatment . . . A plough thus attended will accomplish an acre and quarter in a day, even when the crop is abundant.[53]

Fig. 56. Women from Achill island, Co. Mayo, harvesting potatoes in Ayrshire, Scotland (Dept. of Agric. *Journal* (1901), vol. 2, p.208).

Horatio Townsend commented that this arrangement was extremely well managed, and an improvement on that used for drills, which was similar, except that the labourers were not employed in pairs.[54] The arrangement of two men and pickers following a plough was also recorded by Rawson in county Kildare in 1807, although in this case two boys with light forks were also employed, to 'shake out' the potatoes to the diggers.[55]

Lifting potatoes by hand was dirty and tiring. The labour involved has been well described by several Irish writers. In this century, however, the large-scale use of Irish men and women for harvesting potatoes has not been confined to Ireland, but has been especially associated with Scotland. Squads of 'tattie howkers' went over to Scotland each year for the potato harvest, especially from the west of county Donegal, and Achill Island, county Mayo (Fig. 56). The arrangement of workers in a field has been excellently described by Patrick Macgill:

The way we had to work was this. Nine of the older men dug the potatoes from the ground with short three-pronged graips. The women followed behind, crawling on their knees and dragging two baskets a-piece along with them. Into these baskets they lifted the potatoes thrown out by the men. When the baskets were filled I [a young man] emptied the contents into barrels set in the field for that purpose. . . The first day was wet. . . The job, bad enough for men, was killing for women. All day long, on their hands and knees, they dragged through the slush and rubble of the field. The baskets which they hauled after them were cased in a clay to a

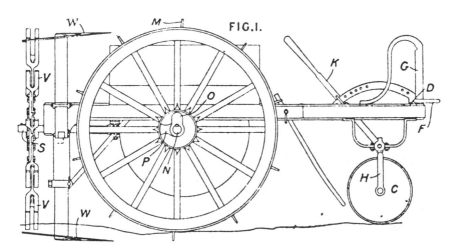

Fig. 57. The mechanical potato digger patented by J. Hanson of Doagh, Co. Antrim, in 1855 (UFTM L2418/6).

depth of several inches, and sometimes when emptied of potatoes a basket weighed over two stone. Pools of water gathered in the hollows of the dress that covered the calves of their legs.[56]

The systematic methods devised for cultivating potatoes in drills meant that harvesting could be potantially more orderly and efficient, although we have seen from Townsend's comments above on county Cork that this was not always the case. However, the even spacing of drills and the relative looseness of the soil meant that the mechanised harvesting of potatoes was made much easier. Fittings were developed in the nineteenth century which could be attached to the front of a plough and which pushed the potatoes out of a drill as it was split open during ploughing. These fittings could be similar to very large paring shares, but more often seem to have been arrangements of iron bars fanning out backwards from the front of the plough. As the plough cut through the drill, soil passed through the bars, but larger tubers were pushed out of the drill.

Experimental potato digging machines were developed in England and Scotland during the early nineteenth century, but it was an Irishman, J. Hanson of Doagh, county Antrim, who patented one of the most successful designs in 1855[57] (Fig. 57). Two horses pulled this machine, and as they did so a broad knife blade, or 'plough piece', cut under the drills. The earth and potatoes undercut by the blade were then scattered by revolving iron forks fitted to the back of the machine. The potatoes were usually prevented from being thrown too far by a net or screen fixed to one side of the forks. Hanson's machine became the prototype for many later designs. The labour of the potato harvest was thereby confined to lifting the potatoes left lying to one side of the drill. The speed with which the machine dug the crop, however, meant that a fairly large labour force was required for a fairly short time. This labour could be hired, but in many areas farmers relied on help from neighbours. A gathering or *meitheal* of neighbours could harvest the entire crop of a small farm in a day, provided the stalks of the plants had been cut and removed beforehand.

Storage of Potatoes

The most common way to store potatoes during the nineteenth and early twentieth centuries was to make a potato pit (also known as a clamp, a bing, or in Irish, *póll-prataí*) (Fig. 58). The term 'pit' is slightly misleading, since most of the potatoes were piled up above ground level. The recommended width of the heap varied between four and six feet, but the pit could be made to any length.[58] The base of the pit was usually dug out to a depth varying between several inches and one foot. After this the potatoes were built up in a gently rounded heap. Martin Doyle advised that each layer of potatoes should be separated by a four to five-inch layer of dry earth, sand or ashes. He also advised that a shallow drain should be dug around the heap.[59] *Purdon's Practical Farmer* suggested a more elaborate construction, with a row of drain tiles or pipes running lengthwise underneath the pit, and vertical ventilation pipes set into the heap at every three or four yards.[60] The heap of potatoes was covered with straw. Some farmers also used rushes or the stalks of the potato plants, but the latter were claimed to injure the crop, since if not properly dried, they would become mouldy and rotten.[61] The pit was completed with a covering of several inches of earth, beaten firm using the back of a shovel. When potatoes were needed during winter, one end of the pit could be opened and tubers removed. The pit was then resealed.

Potato pits, if carefully made, were approved by agriculturalists. In 1814, for example, the Rev. W. Mayne of Dunaghy Parish, county Antrim, wrote that, 'Since I have practised keeping my potatoes in pits in the field, I have never had what is called a curled potato. They are put up close and covered with dry mould only, and when taken out for use,

Fig. 58. Storing potatoes in a 'pit' or 'bing', near Newcastle, Co. Down (WAG 280).

they are as well tasted as when newly turned out of the drill or ridge'.[62] In 1812, however, another cleric in county Cork complained that, 'The practice is lately become too prevalent of storing potatoes in numerous small pits in the potatoe-field, while the operation of digging or ploughing out continues. The farmer is too indolent afterwards to remove them speedily, and the ground is reserved for spring-corn. If this practice should continue, we shall soon see the country entirely divested of wheat-crops'.[63] Apart from this gloomy prediction, however, potato-pits were seen as the most efficient means to store potatoes. The alternative practice, of storing large heaps in a cellar or out-house, was actively discouraged after the 1840s, when it was recognised that these heaps were especially vulnerable to what became known simply as 'the disease'.

Potato Blight and its Treatment

Potato blight (*Phytophthora infestans*) had appeared in America some years before it reached Europe. It was first reported in Ireland in counties Wexford and Waterford in early September, 1845.[64] After a short time the threat it posed to the county became evident. The horrific eye-witness accounts of the sufferings of the Irish people during the

Fig. 59. Experimental spraying of potatoes in Ulster, during the 1930s (UFTM, L1483/7).

following years still make grim reading. At the same time as famine relief was slowly being organised, however, investigations were started to find the cause of the disease. The possibilities discussed have been well summarised by T.P. O'Neill. Some of these were ludicrous, ranging from 'the introduction of guano manure to a curse from God incurred by catholic emancipation and the government grant to Maynooth. Some blamed lightning for blasting the crop. Among the Connaught people it was believed . . . that the northern fairies [had] blighted, or rather carried off the potatoes'.[65] Out of all this, however, two possible causes emerged as the most likely. These were that excessive wetness had rotted the crops, or that the decay was caused by a fungus. For several decades the first of these two possibilities gained most acceptance. The ventilation funnels in pits, advised by Purdon, were one preventive measure advocated in a government leaflet of which 70,000 copies were distributed.[66] Measures such as this, however, proved ineffective in preventing blight. By the 1860s, some agriculturalists were resigned to the idea that the disease was a permanent possibility. 'We need not enter into any speculations relative to the nature and causes of the potato "disease". Volumes have been written on the subject, and as yet, in many respects, we know as little about it as we did in 1845'.[67]

The blight was in fact a fungus, which attacked the leaves of the plant first, and then spread through the foliage. From this it could be washed on to the tubers by rain.[68] The use of copper sulphate, often mixed with

Fig. 60. An early twentieth-century advertisement by Harringtons of Cork, stressing the benefits of spraying potatoes (UFTM, L2455/4).

lime, was well established as a mixture used to kill fungus in wheat and, in Europe, vines. In 1882, an accidental spraying of potatoes beside a French vineyard led to the realisation that the mixture could also kill blight. This led to the development of the Bordeaux mix, and later the Burgundy mix, which was made up of copper sulphate and washing soda. O'Neill has shown that in fact an Irishman, Garrett Hugh Fitzgerald of Courtbrack and Corkaree, county Limerick, had written to the Chief Secretary in Dublin Castle in 1846, informing him of the effects of copper sulphate on blight, but the letter was ignored.[69]

In the last decade of the nineteenth century, a major drive to introduce the spraying of potatoes began in Ireland, especially in the western counties which had been designated as 'congested districts' by the Westminster Parliament in 1891.[70] Spraying machines were developed, both to be carried on the back, or to be pulled by horses (Fig. 59). By 1907, the Irish Department of Agriculture was offering farmers loans to buy horse-drawn spraying machines, and had instituted a scheme through which hand-operated machines could be hired.[71] Through leaflets, placards and advertisements both the Department and commercial firms marketing sprays publicised the benefits of spraying (Fig. 60). The potato crop was sprayed first at the end of June, or early July, and it was recommended that a second spraying should be given three or four weeks later. In 1907, Department inspectors found that while spraying had only been practised to a limited extent in Leinster, in Connaught, Munster and Ulster it was widespread.[72] This may reflect the success of the Congested Districts Board campaign mentioned above. In the following years spraying became standard practice even on small farms.

Fig. 61. Early potatoes, grown on the Lisadell estate, being carted to Sligo town for shipment to Britain (Dept. of Agric. *Journal* (1906), vol. 6, p.9).

Potato cultivation remained of major importance in Ireland for some time after the Famine. There was an initial decline in acreage after the late 1840s, but by 1859, 1,200,247 acres were cultivated.[73] This fell away during the later nineteenth century, and except on small farms in the west, by the 1890s potatoes had lost their dominant position in the Irish diet. The home consumption of potatoes, however, has remained much higher in Ireland than in Britain.

During the present century, the commercial production of potatoes has become highly organised. The cultivation of 'early' potatoes was encouraged by the Department of Agriculture. In 1905, it was reported that although 'There has been no rapid extension . . ., the industry has assumed such proportions as to affect the English and Scotch markets, and this year the fields of growing crops were visited by large numbers of merchants from England and Scotland'.[74] The infrequency of frosts in western Ireland assisted the development of the project. In Clonakilty, county Cork, in 1905, the harvesting of early crops was started on the 3rd of June, and potatoes grown at Lisadell were sent for shipment to Sligo on the 21st of June[75] (Fig. 61).

During the second decade of this century, a successful seed potato industry was started in Ireland which continued to develop in both Northern Ireland and the Irish Free State. R.N. Salaman, who was responsible for developing around twenty varieties for the southern industry, provides a happy ending to the history of potato cultivation in Ireland. In the 1940s, he visited small farms in the west, where by this time the cultivation of seed potatoes was part of a 'well-organised, scientifically supervised and successful industry', and concluded, 'I have found it difficult to recognise in these knowledgeable, intelligent and efficient growers, the direct descendants of those who, less than fifty years ago, had been degraded and exploited by the same potato'.[76]

NOTES

1. Cullen, L.M., *Life in Ireland* (London, 1968), pp.122-123.

2. Mokyr, Joel, *Why Ireland starved* (London, 1983).

3. Salaman, R.N. *The history and social influence of the potato* (Cambridge, 1949 (1970)), p.222.

4. *Ibid.*

5. Salaman, R.N., 'The influence of the potato on the course of Irish history' (The tenth Finlay Memorial Lecture) (Dublin, 1944), p.17.

5. Young, Arthur, *A tour in Ireland*, vol. 1 (Dublin, 1780), p.261.

7. *Ibid.*, vol. 2, 'General observations', pp.33-34.

8. Mitchell, Frank, *The Irish landscape* (London, 1976), p.205.

9. Young, Arthur, *op.cit.*, vol. 2, 'General observations', pp.33-34.

10. MacGill, Patrick, *Children of the dead end* (London, 1914), pp.35-36.

11. Dubordieu, John, *Statistical survey of the county of Antrim* (Dublin, 1812), p.300.

12. Salaman, R.N., *op.cit.* (1949 (1970)), p.288.

13. *Ibid.*, pp.162-168.

14. Interview with Mr. W. Donnelly (Ulster Folk and Transport Museum Tape C77.49)

15. Young, Arthur, *op.cit.*, vol. 2, p.9.

16. *Ibid.*, pp.262-263.

17. De Latocnaye, *A Frenchman's walk through Ireland, 1796-7*, trans. J. Stevenson (Belfast, 1917), p.123.

18. Young, Arthur, *op.cit.*, vol. 2, 'General observations', p.222.

19. Rawson, T.J., *Statistical survey of the county of Kildare* (Dublin, 1807), p.152.

20. Mason, William Shaw, *A statistical account, or parochial survey of Ireland*, vol. 1 (Dublin, 1814), p.263.

21. Young, Arthur, *op.cit.*, vol. 1, p.85.

Irish Farming

22. Gailey, R.A., 'Spade tillage in south-west Ulster and north Connaught', *Tools and tillage*, vol. 1:4 (Copenhagen, 1971), pp.228-229.

23. Bell, J., 'A contribution to the study of cultivation ridges in Ireland', *Journal of the Royal Society of Antiquaries of Ireland*, vol. 114 (Dublin, 1984), p.88.

24. Sampson, G.V., *A memoir, explanatory of the chart and survey of the county of London-Derry, Ireland* (London, 1814), p.304.

25. Interview with Mr. Joe Kane, Drumkeeran, Ederney, County Fermanagh (Ulster Folk and Transport Museum tape R.84.40).

26. Young, Arthur, *op. cit.*, vol. 1, p.75.

27. *The Munster farmer's magazine*, vol. 2 (Cork, 1813), p.109.

28. Young, Arthur, *op. cit.*, vol. 1, pp.73, 79, 83, 169, 209, 215.

29. *Ibid.*, p.33.

30. *Ibid.*, p.73.

31. County agricultural instructors, *Agriculture in Ireland* (Dublin: *The farmer's gazette*, 1907 (?)), p.10.

32. *Ibid.*, p.61.

33. Baldwin, Thomas, *Introduction to practical farming*, 23rd ed. (Dublin, 1893), p.11.

34. Sproule, John, *A treatise on agriculture* (Dublin, 1839), p.300.

35. Young, Arthur, *op. cit.*, vol. 1, p.125.

36. Baker, John Wynn, *A short description and list, with the prices of the instruments of husbandry, made in the factory at Laughlinstown, near Cellbridge in the county of Kildare* (Dublin, 1767), p.7.

37. Young, Arthur, *op. cit.*, vol. 1, p.211.

38. *Ibid.*, vol. 2, p.245.

39. Baker, John Wynn, *Experiments in agriculture, made under the direction of the Right Honorable and Honorable Dublin Society in the year 1767* (Dublin, 1769), p.ix.

40. Tighe, W., *Statistical observations relative to the county of Kilkenny* (Dublin, 1802), pp.220-221.

41. *The Munster farmer's magazine*, vol. 1 (Cork, 1812), p.73.

42. *Ibid.*, vol. 3 (Cork, 1814), p.253.

43. *Ibid.*, vol. 1, p.51.

44. *Ibid.*, p.52.

45. *Ibid.*, pp.52-53.

46. Murphy, Edmund, 'Agricultural report', *The Irish farmer's and gardener's magazine*, vol. 1 (Dublin, 1834), p.556.

47. Bell, J., 'Wooden ploughs from the mountains of Mourne, Ireland,', *Tools and tillage*, vol. 4:1 (Copenhagen, 1980).

48. Buchanan, R.H., 'Calendar customs, part 1', *Ulster folklife*, vol. 8 (Belfast, 1962), p.33.

49. Young, Arthur, *op. cit.*, vol. 1, p.18.

50. *Ibid.*, vol. 2, p.9.

51. *Ibid.*, vol. 2, p.90.

52. *Dublin penny journal*, vol. 2 (Dublin, 1834), p.283.

53. Townsend, H., *Statistical survey of the county of Cork* (Dublin, 1810), pp.656-657.

54. *Ibid.*, p.658.

55. Rawson, T.J., *op. cit.*, p.153.

56. MacGill, Patrick, *op. cit.*, p.75.

57. *Patents for inventions: abridgments of specifications, class 6 agricultural appliances . . . A.D. 1855-1856* (London: H.M.S.O., 1905), p.13.

58. Doyle, Martin, *op. cit.*, p.470; Sproule, John, *op. cit.*, p.304.

59. Doyle, Martin, *op. cit.*, p.470.

60. Purdon, *op. cit.*, p.339.

61. Doyle, Martin, *op. cit.*, p.470.

62. Mason, William Shaw, *op. cit.*, p.260.

63. Cotter, Rev. G.S., 'Advice to the farmers of the county of Cork, on the culture of wheat', *The Munster farmer's magazine*, vol. 1 (Cork, 1812), p.150.

64. O'Neill, Thomas P., 'The scientific investigation of the failure of the potato crop in Ireland, 1845-6', *Irish historical studies*, vol. 5:18 (Dublin, 1946), p.123.

65. *Ibid.*, pp. 125-126.

66. *Ibid.*, p.128.

67. Purdon, *op. cit.*, p.340.

68. O'Neil, Thomas P., *op. cit.*, p.132.

69. *Ibid.*, pp.133-134.

70. Micks, W.L., *An account of . . . the Congested Districts Board for Ireland, from 1891 to 1923* (Dublin, 1925), p.33.

71. Department of agriculture and technical instruction for Ireland, 'Prevention of potato blight', *Journal*, vol. 2 (Dublin: H.M.S.O., 1902), pp.689-690.

72. *Ibid.*, pp.684-689.

73. Salaman, R.N., *op. cit.* (1949 (1970)), p.321.

74. Wallace, M.G., 'Early potato growing', Department of Agriculture and Technical Instruction for Ireland, *Journal*, vol. 6 (Dublin: H.M.S.O., 1906), p.3.

75. *Ibid.*, pp.3 and 9.

76. Salaman, R.N., *op. cit.*, (1949 (1970)), p.332.

CHAPTER 7

Harvesting Hay

Arthur Young described Ireland's grass as some of the finest in the world.[1] Paradoxically, however, this abundance probably hindered the development of harvesting grass as a crop. The belief that there was no need for hay-making in Ireland goes back at least as far as the eighth century when the Venerable Bede wrote that, 'Ireland is far more favoured than Britain by latitude, and by its mild and healthy climate. Snow rarely lies longer than three days, so that there is no need to store hay in summer for winter use or to build stables for beasts'.[2] Hay was certainly made on Norman manorial and monastic estates, where labour services required of *betaghi* or villeins included the mowing of grass.[3] However, it has been suggested that the small number of Irish words relating to hay-making shows the lack of importance of hay in Gaelic Ireland.[4] As late as 1838, one landowner claimed that in parts of west Connaught some people had not known that it was possible to make hay.[5]

Despite this last claim, however, hay-making does seem to have been common throughout Ireland by the later eighteenth century. Arthur Young found that at Gibbstown, Mr. Gerard, 'one of the most considerable farmers in the country', had 100 acres of hay mown each year for his sheep and bullocks,[6] while at the other end of the farming scale tenants with 15-acre farms around Mahon, county Armagh, would mow two of these for hay.[7] Cottars who did not have enough land to grow hay, often bought it.[8] Young also noted the export of hay from Waterford to Norway.[9] The Royal Dublin Society included hay-making as a topic for investigation in the county statistical surveys published in the early nineteenth century. Hay-making was reported in all the counties surveyed, and in county Dublin especially it was found that 'great hay farmers . . . abound'.[10] Pasture and hay crops grew in importance during the nineteenth century, especially when tillage declined after 1850. By 1904 hay was the principal crop in Ireland, after pasture.[11]

Hay could be included in a rotation as an annual crop, or mown from permanent pasture. When grown as an annual, grass seed was often sown along with a grain crop.[12] Agriculturalists advised that the grain should

138

be sown first, the grass seed being spread either just before the field was given a final harrowing, or after the ground had been rolled, in which case the grass seed could be covered using a bush harrow, and the ground rolled again.[13] Grass was also sometimes sown after potatoes,[14] and also along with flax. In 1808, flax was especially recommended as a crop to accompany clover, because the ground would be well prepared for the flax, the growing plants would shade the young clover from the sun, and when the flax was being pulled many weeds would also be uprooted.[15]

Clovers and Italian rye-grass were the most widely recommended hay plants throughout the whole period covered in this book. However, the most common sources of hay-seed seem to have been the sweepings of the lofts of farms or inns.[16] This was condemned, not only because it was unsystematic, but because the seed would contain a large proportion of white 'English' grasses, which were criticised as little better than weeds.[17] Many Irish farmers, however, believed that the mixture of grasses found in permanent pastures produced the most nutritious hay. It was reported in 1802 that in county Meath farmers deliberately tried to ensure the widest possible variety of grasses by collecting portions of seed from neighbours and mixing these with some red and white clover seeds.[18]

Some observers noted that fine crops would grow on land which had been tilled but then simply left to 'throw up the natural herbage'.[19] Arthur Young thought this was probably due to the moist Irish climate: 'The amazing tendency of the soil to grass would prove it if any proof was wanting. Let General Cunningham and Mr. Silver Oliver recollect the instances they shewed me of turnip land, and stubble let without ploughing and yielding the succeeding summer a full crop of hay. These are such facts as we have not an idea of in England'.[20] Early in the present century, most hay made in Ireland was mown from land which had been under grass for a considerable number of years.[21] At the same time, however the specialised production of hay-seed was well established. In county Antrim both perennial rye-grass and Italian rye-grass were grown for seed, and flourishing grass-seed production in counties Cavan, Monaghan, and the north of county Dublin ensured that Ireland did not become a net importer of grass-seed until after the 1960s. Farmers sometimes concentrated on the production of hay-seed to the exclusion of providing hay as fodder for their own stock. One county Derry farmer claimed that in small farms in his area grass-seed was produced for a market held in the town of Dungiven, while stock on the farm 'existed' on oat-straw.[22]

Mowing Hay

Hay was mown in Ireland between July and September, or even October in upland areas and the west.[23] Most agriculturalists were cautious about fixing a date when hay-making should begin, but they were generally agreed that in Ireland mowing was started much too late in the year. Throughout the nineteenth and twentieth centuries there was a general consensus that grasses were best cut just when they were coming into flower.[24] After this, most nutritive matter went into the formation of seeds, while the stalks rapidly became more fibrous.[25] Some loss was accepted as inevitable, however. Where a mixture of grasses was grown, each type came into flower at a slightly different time. Rye-grass, for example, bloomed earlier than clover, so that by the time the latter flowered, some of the nutrition in the rye-grass had been lost.

Arthur Young believed that the delay in cutting hay in Ireland was sometimes due to the late sowing of grass-seed,[26] but in 1824 Hely Dutton alleged that the main reason county Galway farmers left grass uncut in fields they had rented for a year, i.e. in con-acre, was a desire to get as much as possible from the land.[27] A similar motive was identified in a leaflet produced in 1904 by the Irish Department of Agriculture: 'The delay in cutting obviously arises from the desire to secure an increase in the weight of the crop, but while the postponement of the cutting until an advanced stage of ripeness has been reached may ensure a slight increase in the yield, it also entails a serious depreciation in the feeding value of the fodder'.[28] The same leaflet pointed out that not only would early cutting mean that what was lost in quantity would be gained in quality, but that crops would be less weedy, and — most important — there would be a better chance of good weather throughout the hay-making process.[29]

Throughout the nineteenth and early twentieth centuries scythes were widely used for harvesting hay. The general history of scythes in Ireland will be discussed in the chapter dealing with the grain harvest. Early nineteenth-century discussions of the use of scythes for mowing grass were largely confined to the way scythes were used, rather than to a consideration of the efficiency of the implements themselves. In 1824, for example, Hely Dutton railed against the iniquities of scythesmen in county Galway: 'The usual lazy method of mowing is injurious to both [hay and aftergrass] . . . it leaves much of the grass uncut, and the stumps that are left, are worse than useless. When they cut meadow for themselves that they buy standing [on the other hand] they . . . shave it into the earth'.[30] In the mid-nineteenth century, scythes were often

Fig. 62. A Burgess and Keys hay-mowing machine (*The agricultural review* (1860), vol. 1, p.viii).

compared with mowing machines. Combined mowing and reaping machines were being manufactured by American and British farms during the 1850s[31] (Fig. 62). The most popular early machines seem to have been those manufactured by the firms of Woods, and Burgess and Key. The correspondence columns of the *Irish farmer's gazette* during the 1860s include some worried queries about the possibility of Woods machines becoming clogged, especially in bottom meadows where grass was soft and lush.[32] Generally, however, the machines were welcomed enthusiastically. In July 1863, a Woods machine was demonstrated near Ballina, county Mayo:

> Machines of that description have not hitherto been introduced into that part of Connaught, and the recent trial was, therefore, quite a novelty in this neighbourhood. The machine was put into the hands of a man who was quite inexperienced in its management, notwithstanding which circumstance a meadow of unusually heavy rye-grass and clover was mown in splendid style, at the rate, we are informed, of about an Irish acre per hour. The machine was purchased by Mr. Pery [the farmer on whose land the demonstration had taken place], and other gentlemen in that quarter have also sent in their orders for machines of the same kind.[33]

One reason given for the spread of mowing machines during the 1860s was the difficulty of obtaining scythesmen, 'at least in sufficient numbers

Fig. 63. Harvesting grass for seed in Ulster, using a reaper-binder, C.1920 (WAG 1977).

and at anything like sufficient wages'.[34] In 1863, *Purdon's practical farmer* estimated the comparative costs of mowing a field using a Burgess and Key machine, and using scythes. Interestingly, it was assumed that the machine and its operator could be hired for the work, while the farmer would provide the horses. The cost of this — three shillings per Irish acre, or about one shilling and ten pence per statute acre — was claimed to be much cheaper than hiring scythesmen, which was estimated to cost seven shillings per Irish acre, or four shillings and four pence per statute acre.[35]

Irish farmers were encouraged by state institutions to buy mowing machines. In 1904, the Department of Agriculture and Technical Instruction recommended the small one-horse mowing machines manufactured by Pierce of Wexford, and the Wexford Engineering Company. Small farmers who did not think that the size of their holdings warranted the outlay of capital required to buy a mower were urged to combine with other small farmers to buy the machines co-operatively.[36] Loans were also given to help with the purchase. In July 1912, a scheme was started in western 'congested' counties to provide loans for the purchase of a range of implements costing £6 and over. By September 1912, 110 mowers had been purchased under the scheme.[37] At the same time, however, it was recognised that in some circumstances scythes could still be used more efficiently than mowing machines. An English

Board of Agriculture leaflet, distributed by the Irish Department in 1903, recommended that red clover hay was better cut with a scythe, because a better aftergrowth resulted. One explanation suggested for this was that scythes had a more oblique cutting action than mowing machines, and this meant that rain tended to run off the stubble, rather than enter the stalks and rot the roots. The Board leaflet concluded, however, that the better aftergrowth was probably due more to the longer stubble left by the scythe.[38]

By the early years of this century, when rye-grass was being grown for seed, it was sometimes cut using a reaper-binder (Fig. 63). In this case the grass was handled in much the same way as a cereal crop. Four or six sheaves were placed against one another to form a stook. After several days the stooks were collected into 'rickles' or 'huts' composed of twelve to sixteen stooks. In eastern Ulster, threshing was often carried out after the large stacks had been constructed.[39]

Hay-Making

The main criticisms of Irish hay-making techniques, apart from late mowing, were that the hay was turned too often, and left in the fields too long. Arthur Young's comments on hay-making near Lough Derg on the Shannon were echoed frequently during the next two centuries: 'They shut up their meadows for hay in March or April, and rarely begin to mow till September. I should remark that I saw hay-making or marring all the way (October 3rd) from Johnstown hither, with many little fields covered with water, and the cocks forming little islands in them. They are generally two months making it'.[40] Agriculturalists generally agreed with Thomas Baldwin who wrote in 1874 that, 'The ruling idea in the saving of hay should be to allow it to remain in the field as short a time as is absolutely necessary... Well-made hay retains its green colour, and possesses an agreeable odour'.[41]

The first stage in hay-making everywhere in Ireland was to turn or 'shake-out' the new-mown hay. This was particularly important when scythes had been used, as they left the cut grass lying in heavy swathes.[42] Early mowing machines spread the grass more as they cut it. In 1844, it was recommended that four or five hay-makers, who could include 'women and boys', should follow each scythesman. The assumption that women would be part of a hay-making work team makes an isolated, rather sneering comment by Arthur Young that 'Irish ladies

Fig. 64. A Smith and Ashby hay-making machine, which took the prize at a trial of haymakers in Derry, in 1858 (*The agricultural review* (1860), vol. 1, p.cxlix).

in the cabbins cannot be persuaded, on any consideration... to make hay, it not being the custom of the country'[43] more intriguing, but also more dubious. Shaking out the hay was often done with forks, although at least one early nineteenth-century commentator condemned this as lazy, and recommended that hay-makers should use their hands instead.[44] Horse-drawn hay tedders, used for turning and spreading hay to dry, were developed in Britain in the early nineteenth century. A tedder designed by Robert Salmon of Woburn became the prototype for many later machines (Fig. 64):

> It was an axle on a pair of wheels. The axle shaft carried two smaller wheels, the rims of which were joined at regular intervals by horizontal bars, making a hollow cylinder. The horizontal bars carried curved iron teeth, pointing outwards about six inches long. This arrangement revolved as the machine was drawn along, and the series of forks scattered the hay in all directions, and over the driver and the horse as well, if care was not taken to drive to leeward.[45]

By the 1860s, hay tedding machines were well-known in Ireland, where their introduction was considered 'a great boon to farmers'. Some experts, however, thought the machines turned hay too roughly. Clover hay especially was thought to suffer, as it became brittle when drying, and the finer leaves were liable to be lost if the hay was shaken.[46] But

Fig. 65. A horse-drawn hay rake in use on an Ulster farm c.1920 (WAG 3371).

whether tedders were used or not, it was alleged that, in general, hay in Ireland was turned too much. A description of hay-making practices in county Dublin in 1802 was particularly bleak: 'The hay is in general turned, and turned over and over again, until at length it is frequently caught by rain; then it must be dried, and this process of turning is renewed and continued, until it is devoid of either colour or smell'.[47]

Mid-nineteenth century texts advised that hay which had been cut and tedded in the morning should be raked into long 'wind-rows' in the afternoon. This could be done using hand-rakes, or with horse-drawn rakes, which were in use in England by 1831. The horse rakes which gained most common acceptance were constructed on a frame set on two wheels. A row of curved teeth between the wheels collected hay until its own weight caused it to fall over (Fig. 65). The slightly rounded teeth meant that the falling bundle of hay was made into a loose roll, which was praised because it allowed a free circulation of air through the mass. These rolls were sometimes likened to ladies' muffs, a description also applied to 'lap-cocks' which will be described below.[48]

If the weather was very fine, the next stage in hay-making could be carried out in the evening of the day on which the hay had been cut, tedded and raked. This involved building the hay into small heaps known as field cocks. In 1839, it was suggested that this work could be carried out by boys and girls under the supervision of the farmer or a

Fig. 66. A field of lap-cocks in the west of Ireland (Johnson, C. (1901), p.222).

'confidential servant'. The young people were split into small groups which worked along sections of the field made up of either three or five ridges. Each group was divided into carriers and rakers. The carriers gathered the hay on to the central ridge, on which a row of cocks was to be built. A raker followed each carrier, collecting any hay left behind. It was suggested that where possible five people should work on each row of cocks; a carrier and a raker on each side of the row, and a person (preferably a more experienced worker) to build the cocks.[49]

The technique of building the smallest field cocks shows a broad regional variation in Ireland between north and south. Throughout most of the southern half of the country the cocks were simply small 'bee-hive shaped' heaps of hay, but in the north (including counties Leitrim,[50] Mayo,[51] and Meath[52]), lap-cocks (Irish: *gráinneog*) were made (Fig. 66). These were warmly approved of by agricultural writers. The following technique of making a lap-cock was given in a nineteenth-century English household guide, where Irish lap-cocks were recommended to English farmers in wet seasons (Fig 67):

> The rakers follow each other, raking about four feet in breadth into the spread grass. A laper, generally a woman, follows each rake, collecting a small quantity of hay evenly between her hands, which she places upon the raked round, doubling in one end as one presses it down, and with the right

Fig. 67. Mr. Paddy Cassidy of Lisnaskea, Co. Fermanagh, demonstrating how to make a lap-cock (UFTM, L2224/2/8).

or left hand doubles in the other end, placing a small handful of loose hay across on the top.[53]

These rolls of grass, being open at the centre, did not pack together, and air could circulate and dry the hay evenly, and quickly. It was claimed that because of their construction, lap-cocks had been known to survive eight days of almost incessant rain without the hay being damaged.[54] Lap-cocks and beehive-shaped cocks were usually spread out and remade several times over a period which could last anything between two days and more than a week.[55] Agriculturalists were united in agreeing that this was much too long.

The next stage in hay-making involved combining two or more small cocks to make bigger ones.[56] In some areas this intermediate process of cocking or ricking was eliminated.[57] Elsewhere, however, the larger cocks were spread out and remade until the hay was considered completely dry, so that this stage of the process could also last for a week

Fig. 68. A tumblin' Paddy in use, c.1955 (photograph courtesy of *The Irish Times*).

or more. After this the cocks were again combined to make ricks or 'trampcocks', co-called because someone stood on the cock as it was being built, to compress the hay. Trampcocks varied in size, anything between 10 cwts. and a ton of hay being built into a single cock.[58]

By the beginning of the nineteenth century, some farmers had adopted methods which assisted the moving of larger amounts of hay collected in the later stages of hay-making. One practice, which seems to have been fairly common by 1800, was to put a rope around the back of a cock which was to be moved, and then to fix each end of the rope to a horse, either by means of a swing-tree, or directly to the horse's harness traces. This technique meant that two or more smaller cocks could be dragged along the ground at once.[59] In 1802, one large farmer in county Meath extended this method so that even the largest ricks could be moved. He replaced the rope by a chain, and the draught was provided by six oxen.[60] Later in the century hay sweeps or collectors, developed in America and Scotland, became popular. Both American and Scottish sweeps were designed so that when full of hay they could be easily tipped over and the hay deposited in a heap. The sweeps which became known as 'Tumblin' Tams' in Scotland and 'Tumblin' Paddies' in Ireland were particularly common (Fig. 68). These consisted of a row of six tines, each about three feet in length, spaced along a beam at intervals of eighteen inches. Two handles joined to the beam allowed the driver to manipulate the sweep as it gathered up hay. A horse pulled the sweep with draught chains which were attached to swivel links at each end of the beam. The chains helped to support the hay being collected, while the swivels facilitated unloading. This was achieved by allowing the points of the tines to catch in the ground, so making the sweep tip forward and turn completely over, releasing its load as it turned. When a full revolution had taken place, the driver took hold of the handles again and began to collect another load. Sweeps could carry almost half a cart load of hay.[61]

The tardiness of Irish hay-making practices was especially criticised when the length of time between making trampcocks, and making these into large permanent stacks, was discussed. Farmers were accused of developing a false sense of security once the largest field cocks had been made. Hay could be left in these ricks for weeks, or even months, by which time their bottoms, tops and sides were often completely rotted.[62] In 1802, Hely Dutton alleged that the great hay farmers of county Dublin often sold this ruined hay to unsuspecting buyers at Dublin's Smithfield market:

> all the bad and refuse hay of the farm . . . is lapped up inside of each roll of hay, and loaded for Smithfield. The frauds, that are practised in loading hay for this market call loudly for redress; it is a very common practice with many hay-farmers to shake a little fine hay on the ground; then a quantity of bad is shook evenly over this, and lapped up and loaded for market. In the market the farmer's man stands with a handful of hay, drawn from some part that has not been doctored.[63]

Hay which was kept for use on the farm could be stacked in a yard or haggard, or simply in the corner of a field. In the later nineteenth century, hay was often brought to the stacks on rick-shifters (Fig. 69). These were low carts with a large flat body which could be tilted backwards. Field cocks or ricks were then pulled up on to the cart, being drawn by a rope or chain tied around their backs in a way similar to that which had been used earlier in the century to drag smaller cocks along the ground. The ricks were pulled on to the shifter, not by a horse, however, but manually, using a winching mechanism fixed to the front of the cart.

Hay stacks could be built on either a circular or an oval base. The stack could be raised from the ground simply by spreading a layer of straw or sticks on the ground, but on even some small farms stone stack stands were also constructed. The stack was made by spreading the hay

Fig. 69. Hay being carried on a ruck-shifter, c.1955 (photograph courtesy of *The Irish Times*).

regularly, but keeping the centre well raised. When the hay had been built up to a height of several feet, either with parallel sides or so that they bulged slightly outwards, the roof was begun. This was conical and steep. After the form of the stack was complete, loose hay was pulled off the sides to make the surface more smooth and compact. The completed stack was thatched with straw and tied down with straw or hay ropes.[64] Nineteenth-century farming journals contain many advertisements for rick covers (Fig. 70) which gave additional protection to the hay. By the 1840s cloths were said to be 'much in use at gentlemen's places, but amongst the farming class they are still a rarity. No farmer should be without one, as in one season he might save more by having it than twice the first cost. By having one of these cloths a farmer can bring in his hay and rick it whenever fit to be made into tramp-cocks in the meadow'.[65]

Stacks could be damaged by damp which led to heating, over-

Fig. 70. A stack cover (*The agricultural review* (1860), vol. 1, p.cxlix).

fermentation and eventual rotting of the hay. However, it was frequently claimed that these processes could at least be hindered by mixing salt with the hay as the stack was being built. The amount most generally advised was one stone of salt for every ton of hay,[66] although *Purdon's practical farmer* suggested that up to two stones per ton could be used.[67] Salting was not universally advised, however. Martin Doyle argued that hay given to horses should not be salted, since this would increase their thirst and hence their intake of water to undesirable levels.[68] In 1823, the North West of Ireland Society *Magazine* claimed that, far from preventing dampness, salting hindered drying, as it would actually attract moisture during wet weather.[69] This seems contrary to generally held views, however.

By the early twentieth century, hay barns were widely in use. Many of these were iron or wooden-framed structures, covered with sheets of corrugated iron. State institutions such as the Department of Agriculture and the Congested Districts Board gave financial assistance to farmers to build these barns.[70] This policy was so successful that as one British text commented, hay barns were 'nowhere in this country so fully appreciated as in the Emerald Isle'.[71]

Many agriculturalists were insistent that Irish hay-making practices revealed a total lack of concern for the loss of nutritional value of the crop when it stood in the field exposed to the bleaching and withering action of rain, sun and wind. Rain especially was seen as harmful, not only because it accelerated rotting, but also because it washed soluble nutrients out of the crop.[72] In 1874, Thomas Baldwin calculated that losses during hay-making in Ireland amounted to 20 per cent of the total value of the crop. Baldwin estimated that there were about 1,500,000 acres of meadow in Ireland, which produced about 2 tons per acre, or 3 million tons in all. Taking the current value of hay to be at least 50 shillings a ton, he calculated the total potential value of the hay crop to be £7,500,000. Of this, he claimed, £1,500,000 was lost by bad management.[73]

As with many Irish farming practices which were condemned by agriculturalists, the farmers themselves could defend their methods. Their central aim throughout the hay-making process was to prevent over-heating, and therefore over-fermentation of the crop. In 1808, it was reported that in county Clare, if a farmer put his hand into a cock and found it in the least warm, the hay would be immediately spread out again.[74] Agriculturalists argued that gentle heating of hay could in fact be beneficial, as the fermentation which this produced converted starch in the hay into sugar. In some areas in England, farmers allowed heating to continue until the hay had turned brown, and acquired a 'strong burnt-like smell'.[75] Some Irish agriculturalists conceded that this was going too far. Hely Dutton thought it an example of 'English obstinacy'.[76] Over-fermentation meant that the sugar content of the hay could be converted first in to alcohol, and then into acetic acid.[77] Extreme over-heating could lead to spontaneous combustion in stacks. Irish farmers, defending their attempts to prevent hay from heat or fermenting at all, claimed that the ill effects of these processes were more likely to happen in Ireland than in England since Irish grass was more luxuriant. In 1802, Thompson recorded the case of a former Bishop of Meath which was cited as evidence for this. The bishop 'pursued the English mode of hay-making, the consequence of which was, that about sixty ton of his hay took fire in Ardraccan haggard.'[78] An even more spectacular combustion was described in *The farmer's gazette* of 1847:

> We remember once seeing 500 tons of hay all but consumed by spontaneous combustion, on account of the new-fangled economy of a steward, who fancied he would conquer the elements by which nature surrounded him, nor did he fail without loss of life, for on lifting off the tarpaulin cloth by which his burning hay-rick was covered, two men were deprived of life.[79]

The conclusion drawn in this chapter is one which appears frequently in texts in which Irish writers responded to the criticisms of British agriculturalists. While lamenting the carelessness of much Irish hay-making; the writer quoted below makes a strong attack on rigid theoretical improvers:

> Very many who endeavour to effect reforms in our husbandry think they can best do so by holding up the examples of our excellent friends at the other side [Britain]. This, no doubt, is plausible and praiseworthy, provided it were done with due regard to the difference in the circumstances of soil, climate, etc., of the three countries; but these differences were never taken into account by the men brought into this country for its enlightenment, and although they did much good in introducing any change, for none could be for the worse, they have latterly proved a sad drag-chain on improvement, as in point of prejudice against everything new or unknown to themselves, they are "Ipsis Herbernis Hiberniores".[80]

NOTES

1. Young, Arthur, *A tour in Ireland*, vol. 2, 'General observations' (Dublin 1780), p.140.

2. Bede, *A history of the English church and people* trans. Leo Sherley-price, rev. ed. R.E. Latham (Harmondsworth: Penguin, 1983), p.39.

3. Lane, Padraig, *Ireland* (London, 1974), p.18.

4. Adams, G.B., 'Work and words for haymaking', *Ulster folklife*, vol. 12 (Holywood, 1966), p.69.

5. [Devon Commission], *Digest of evidence taken before Her Majesty's Commissioners of inquiry into the state of the law and practice in respect to the occupation of land in Ireland*, vol. 1 (Dublin, 1847), pp.635-636.

6. Young, Arthur, *op. cit.*, vol. 1, p.51.

7. *Ibid.*, p.172.

8. Young, Arthur, *op. cit.*, vol. 2, 'General observations', p.27.

9. *Ibid.*, p.195.

10. Dutton, Hely, *Observations on Mr. Archer's statistical survey of the county of Dublin* (Dublin, 1802), p.74.

11. Department of Agriculture and Technical Instruction for Ireland, 'Haymaking', Leaflet no. 46, *Journal*, vol. 4 (Dublin: H.M.S.O., 1904), p.716.

12. Sproule, John, *A treatise on agriculture* (Dublin, 1839), p.415; Baldwin, Thomas, *Introduction to Irish farming* (London, 1874), p.33) Wright, R. Patrick (ed.), *The standard cyclopedia of modern agriculture and rural economy*, vol. 6 (London, 1908), p.234.

13. Baldwin, Thomas, *op. cit.*, p.38.

14. Young, Arthur, *op. cit.*, vol. 1, p.219.

F

15. Dutton, Hely, *Statistical survey of the county of Clare* (Dublin, 1808), p.123.

16. Thompson, R., *Statistical survey of the county of Meath* (Dublin, 1802), p.213); Dutton, Hely, *op. cit.*, 1808, p.123.

17. Baldwin, Thomas, *op. cit.*, p.37.

18. Thompson, R., *op. cit.*, p.213.

19. Coote, C., *Statistical survey of the county of Cavan* (Dublin, 1802), pp.207-208.

20. Young, Arthur, *op. cit.*, vol. 2, 'General observations', p.139.

21. Department of Agriculture and Technical Instruction for Ireland, *op. cit.*, p.716.

22. Wright, R. Patrick (ed.), *op. cit.*, p.236 Other information from Dr. Austin O'Sullivan of the Irish Agricultural Museum, Wexford, and Mr. James O'Kane of Claudy, Co. Derry.

23. Young, Arthur, *op. cit.*, vol. 2, pp.210, 246 and 264; Thompson, R., *op. cit.*, pp.215-216; Dutton, Hely, *op. cit.*, 1808, p.129; MacNeill, M., *Lughnasa* (London, 1962), p.61.

24. Sproule, John, *op. cit.*, p.419; *Purdon's practical farmer* (Dublin, 1863), p.355; Wright, R. Patrick (ed.), *op. cit.*, p.419.

25. Sproule, John, *op. cit.*, p.419.

26. Young, Arthur, *op. cit.*, vol. 2, pp.219-220.

27. Dutton, Hely, *A statistical and agricultural survey of the county of Galway* (Dublin, 1824), p.136.

28. Department of Agriculture and Technical Instruction for Ireland, *op. cit.*, p.716.

29. *Ibid.*

30. Dutton, Hely, *op. cit.*, 1824, pp.137-138.

31. Fussell, G.E., *The farmer's tools, 1500-1900* (London, 1952), p.145.

32. *Irish farmer's gazette* (Dublin, 1863), pp.223 and 249.

33. *Ibid.*, p.242.

34. Purdon, *op. cit.*, pp.356-357.

35. *Ibid.*, p.357.

36. Department of Agriculture and Technical Instruction for Ireland, *op. cit.*, p.717.

37. Department of Agriculture and Technical Instruction for Ireland, *Twelfth annual report, 1911-1912* (London: H.M.S.O., 1913), p.45.

38. Board of Agriculture (England), 'Haymaking', Leaflet no. 85: in Department of Agriculture and Technical Instruction for Ireland *Journal*, vol. 3 (Dublin: H.M.S.O., 1903), p.724.

39. Wright, R. Patrick (ed.), *op. cit.*, p.237.

40. Young, Arthur, *op. cit.*, vol. 2, p.246.

41. Baldwin, Thomas, *op. cit.*, p.39.

42. Sproule, John, *op. cit.*, p.419.

43. Doyle, Martin, *A cyclopaedia of practical husbandry*, rev. ed. W. Rham (London, 1844), p.274; Young, Arthur, *op. cit.*, vol. 2, p.140.

44. Dutton, Hely, *op. cit.*, (1808), p.128.

45. Fussell, G.E., *op. cit.*, p.141.
46. Purdon, *op. cit.*, pp.358-359.
47. Dutton, Hely, *op. cit.*, (1802), p.74.
48. Fussell, G.E., *op. cit.*, pp.140-141.
49. Sproule, John, *op. cit.*, pp.419-420.
50. McParlan, James, *Statistical survey of the county of Leitrim* (Dublin, 1802), p.42.
51. McParlan, James, *Statistical survey of the county of Mayo* (Dublin, 1802), p.56.
52. Thompson, R., *op. cit.*, p.215.
53. *Cassell's household guide*, vol. 4 (London, 18?), p.13.
54. Purdon, *op. cit.*, pp.359-361.
55. Tighe, W., *Statistical observations relative to the county of Kilkenny* (Dublin, 1802), p.378.
56. McParlan, James, *Statistical survey of the county of Donegal* (Dublin, 1802), p.56.
57. Tighe, W., *op. cit.*, pp.378-379; Coote, C., *op. cit.*, p.79; Purdon, *op. cit.*, pp.360-361.
58. Tighe, W., *op. cit.*, p.380; Wright, R. Patrick (ed.), *op. cit.*, p.240.
59. Sproule, John, *op. cit.*, p.420; Thompson, R., *op. cit.*, p.217.
60. Thompson, R., *op. cit.*, p.217.
61. Partridge, Michael, *Farm tools through the ages* (Reading, 1973), pp.147-148.
62. Dutton, Hely, *op. cit.*, 1802, p.77; Coote, C., *op. cit.*, pp.79-80.
63. Dutton, Hely, *op. cit.*, (1802), p.77.
64. Sproule, John, *op. cit.*, p.421.
65. *The farmer's gazette*, vol. 7 (Dublin, 1847), p.137.
66. Sproule, John, *op. cit.*, p.421; Doyle, Martin, *op. cit.*, pp.280-281; Department of Agriculture and Technical Instruction for Ireland, *op. cit.*, (1904), p.718.
67. Purdon, *op. cit.*, p.363.
68. Doyle, Martin, *op. cit.*, p.281.
69. North West of Ireland Society *Magazine*, vol. 1 (Derry, 1823), p.267.
70. Wright, R. Patrick (ed.), *op. cit.*, p.245.
71. *Stephens' book of the farm*, vol. 2, 5th ed. rev. James Macdonald (Edinburgh, 1908), p.277.
72. Wright, R. Patrick (ed.), *op. cit.*, pp.232-233.
73. Baldwin, Thomas, *op. cit.*, p.42.
74. Dutton, Hely, *op. cit.*, 1808, p.125.
75. Purdon, *op. cit.*, p.364.
76. Dutton, Hely, *op. cit.*, (1808), p.126.
77. Wright, R. Patrick (ed.), *op. cit.*, p.240.
78. Thompson, R. *op. cit.*, p.215.
79. *The farmer's gazette*, vol. 7 (Dublin, 1847), pp.83-84.
80. *Ibid.*, p.84.

CHAPTER 8

The Cultivation of Flax

Flax, unlike potatoes, and to some extent hay, has had a very long history in Ireland. Linen production is detailed in the Brehon laws, and linen clothes and church vestments are commonly referred to in the literature of the early Christian period. In the later eighteenth century, some flax production was part of the small farm subsistence economy. Arthur Young found at Packenham in Westmeath that, 'The cottars all sow flax on bits of land, and dress and spin it, and it is woven in the county for their own use, besides selling some yarn', while at Hampton, county Dublin, they 'sow enough [flax] for their own use, not enough for manufacture'.[1] During the last three centuries, however, flax cultivation has been closely linked to the large-scale commercial production of linen, and this has made the crop distinctive in several ways. By far the greatest proportion of the crop was grown for the market. Given its commercial importance, it is not surprising that, as John Sproule claimed in 1839, 'The cultivation of flax in this country has engaged a considerable proportion of the attention of the legislature as well as the farmer, and more laws have been framed relating to it than to any other branch of husbandry'.[2]

From early in the eighteenth century, state institutions attempted to organise and develop flax growing. In 1710, a Board of Trustees of the Linen and Hempen Manufactures was established, and some of the earliest premiums granted by the Royal Dublin Society were given for flax, and flax-seed.[3] Nineteenth-century bodies included the Flax Improvement Society, formed in 1841, and the Flax Extension Society, formed in 1867.[4] At the end of the century the Irish co-operative movement encouraged the development of local flax societies, and in 1900 also organised a conference in Belfast for flax growers, scutchers and spinners.[5]

Despite all this institutional activity, however, flax production followed the pattern which can be traced for most crops in Ireland during the last 250 years: an initial period of expansion followed by a slow decline. By the end of the first decade of the nineteenth century about 70,000 acres of flax were grown annually, and this rose to almost

175,000 acres in the early 1850s. The American civil war led to an almost complete cut-off of cotton imports, and the linen industry boomed. In 1864, 301,693 statute acres of flax were grown in Ireland.[6] This was an exceptional period, however, temporarily reversing a decline which John Sproule claimed was beginning in 1839.[7] By 1900, there were less than 50,000 acres of flax grown, and further decline was only reversed during the years of the First World War. Reasons suggested for the decline included competition from exports of cheap Russian flax, the cost of labour, and the reluctance of farmers to commit themselves to the heavy work of harvesting flax so near to the grain harvest. It was also suggested that the use of poor seed had led to a succession of disappointing crops.[8]

Within this overall pattern, one of the most remarkable developments was the increasing concentration of the crop in the province of Ulster. More flax was grown in Ulster than in the other provinces as early as the late seventeenth century,[9] and by 1809 the difference in acreages had become very marked indeed:

Province	*Acres (Irish) of Flax Sown*[10]
Ulster	62,441
Munster	3,716
Leinster	4,107
Connaught	6,485

The decline of flax growing outside Ulster was spectacular during the later nineteenth century. Between 1869 and 1871, the acreage fell by almost a half, and by 1890 there were only 886 acres of flax grown in Leinster, Munster and Connaught combined.[11]

National organisations made periodic attempts to extend flax cultivation outside Ulster, both by giving grants and sending instructors to areas which they considered had most potential. A lot of this activity was centred on Munster, where markets were organised at which northern buyers could purchase locally grown flax. These failed, however, and seem to have been marked more by suspicion and disputes than by awakening enthusiasm. In 1891, for example, 'an unfortunate dispute between the local Scutch Mill owners as to the relative claims of Clonakilty, Dunmanway and Ballineen to have a flax market was the means of preventing Northern buyers attending the southern market.'[12] Contemporaries and historians have explained the failure of flax production in southern Ireland by the lack of markets and local scutching mills. The concentration of industrialised linen production in the north-east was obviously also of central importance.[13] One less serious explanation for the absence of flax cultivation throughout most of Ireland was that farmers thought that the work involved was demeaning, especially the retting or soaking of flax. In 1774, Hyndman quoted, in pidgin Irish, a saying whose sense derived from this attitude:

> Neal fiss egum nach mhaphushe an lenn
> I do not know but he would water (or drown) flax.[14]

Within Ulster during the eighteenth century, linen production became concentrated in areas where there was a constant supply of strongly flowing water for bleaching work, such as the Lagan, Upper Bann, Roe and Faughan Valleys, on the Callan river near Armagh, and around Ballybay in County Monaghan.[15] Before the nineteenth-century industrialisation of linen, production of the fabric had important effects on farming life in general. A dual economy of farming and of spinning and weaving linen developed, and gave the weaver-farmers of the major linen-producing areas a relative economic independence. This led to a lifestyle which Arthur Young considered characterised by insolence and indolence:

> They [keep] . . . packs of hounds, every man one, and joining, they hunt hares: a pack of hounds is never heard, but all the weavers leave their looms, and away they go after them by hundreds. This much amazed me, but assured it was very common. They are in general apt to be licentious and disorderly; but they are reckoned to be rather oppressed by county cesses for roads, etc., which are not of general use . . . They are in general very bad farmers, being but of the second attention, and it has a bad effect on them, stiffening their fingers and hands so that they do not return to their work so well as they left it.[16]

The linen industry also tended to lead to a concentration of tiny farms.[17] The dual economy which weavers engaged in meant that they could only work a small holding, and many hired labourers even for this. In 1839, Sproule argued that flax cultivation was best suited to small farms. The amount of labour required at each stage of cultivating and processing the crop meant, he claimed, that flax would not be profitable when cultivated on a large scale. However, on small farms and cottars' holdings 'the number of hands is large in proportion to the extent of land which is occupied, and more time can, accordingly, be devoted to the culture and management of the crop.'[18]

The Preparation of Land and Planting Flax

Flax grows best on a rich friable loam, but it can also grow well on clay soils, and fairly well on sandy, peaty soils. Arthur Young dismissed a notion which he claimed was common in Northern Ireland, that very rich land was not good for the crop.[19] However, this view was repeated in some later standard farming textbooks, writers arguing that rich land made the plants more likely to lodge, or produce subsidiary stalks as well as the central stem. The smaller stalks were of very little value for fibre. For the same reason it was advised that farm yard manure should not be applied directly to flax ground, but should be used on the crop preceding it in the rotation.[20]

Most agricultural writers agreed that flax could be included at almost any stage in a crop rotation, provided that it was not the first crop to be sown on lea ground, which might not only be too rich, but also might be infested with destructive leather-jacket grubs. Arthur Young claimed that potatoes or turnips were the best preparation for flax,[21] while some nineteenth-century writers advised that it should be put in after oats grown on lea ground.[22] One man interviewed recently near Dromore, County Down, was emphatic that barley grown on lea ground was the best preparation.[23] Whatever its place in the rotation, however, it was generally accepted that flax should not be grown on the same piece of ground very frequently. The gap advised varied between five and ten years.

Flax should be sown on a seed-bed of fine, firm tilth. Agriculturalists described several methods by which this could be achieved. In the eighteenth century, it seems to have been common practice to plough the ground several times. In 1774, Hyndman claimed that the more often the land was ploughed, the better, but that it should be done at least three times. One ploughing did not reduce the land to a proper tilth, and the second brought up the weeds buried by the first. A third ploughing, however, buried these again and broke the ground up sufficiently fine.[24] Nineteenth-century writers often advised one deep ploughing, preferably in later autumn, but at least two months before the land was sown. The ground was finely broken up just before sowing, using a 'strong, coarse harrow', and then rolled. If the land was in ridges, it was recommended that the harrows should be pulled across these to prevent the harrow tines creating runnels along the length of the ridges in which water could lodge.[25] Large stones could also be pulled off the surface by the harrows, and deposited either in the furrows between ridges, or at the edge of the field, where they could be collected afterwards. In eighteenth-century Ulster, however, women and children were often employed to walk or crawl over the field collecting stones. One farmer near Warringstown, County Down, estimated that it would take one woman four days to remove the stones from an acre.[26] Large clods were also broken up with a spade or a mell. The North-West of Ireland *Magazine* suggested that this should be done prior to harrowing, to prevent the clods being gathered into heaps, which would make the seed grow unevenly.[27] Hyndman stressed that clods should be broken up at least before sowing, because breaking them up afterwards would displace seeds which should be spread as evenly as possible.[28] Most writers also advised that the ground should be rolled before sowing, or that if a roller was not available, that a heavy harrow, with its tines turned upwards, should be dragged across it.[29]

Flax should be sown between late March and late April. Hyndman advised sowing the crop as late as possible because in Ireland early sown seed was in danger of being ruined by 'surfeits of rain'.[30] An experienced county Down cultivator who has recently started growing flax again agreed with this, saying that in the past he had sown flax as early as 10th March, but even when the weather was fine, seed sown in mid-April had produced crops of the same quality.[31]

Some eighteenth-century Irish farmers grew their own flax-seed. Arthur Young recorded that, near Armagh, flax intended for coarse linen was allowed to stand unpulled until the seed bolls ripened.[32] For finer linens, farmers preferred American seed for 'light and mountain' land, but Young found that at the time of his travels, this source had temporarily been cut off by the American War of Indpendence.[33] On heavy, clay land farmers preferred Riga Dutch or Flanders seed, claiming that it produced more flax of better quality.[34] In the later nineteenth century, the term Riga was applied to Russian seed, and this and Dutch seed were the two most common types grown. In 1904, Russia and Holland were claimed to provide almost all of the seed used for flax crops in Ireland.[35] In 1908, however, Irish farmers were urged to buy seed only from 'reliable' importers in Belfast or Derry. Some unscruplulous Dutch merchants, it was alleged, were exporting 'a white-flowered . . . type from the province of Friesland, [which] produces short, coarse, much branched stems, and produces inferior crops of poor-quality fibre and a large quantity of seed'.[36] Generally, however, both farmers and agriculturalists agreed that seed saved from Irish flax was not so good as that grown in a less moist climate. Arthur Young found that near Armagh, farmers would sow 2½ to 3 bushels of

Fig. 71. Sowing flax-seed broad-cast from a sheet, near Toome, Co. Antrim, c.1920 (WAG 1019).

imported flax-seed per acre, but when using seed which they had grown themselves, they sowed 4 bushels per acre, to compensate for what they believed was the latter's poorer quality; [37] 2½ to 3 bushels remained the amounts of seed recommended per acre throughout the nineteenth century, but John Sproule recommended that an extra quarter of a bushel should be added as this would produce thick crops with long fine stalks, and few side branches to the main stem. Partly because he believed that flax should be sown as thickly as the quality of the ground allowed, Sproule discouraged the practice of sowing grass-seed along with flax, arguing that flax plants should be grown more closely together than grain crops, and would therefore allow little space for grass to flourish. [38]

Discussions on the best method of sowing flax did not lead to the same controversy as those arising when other crops were discussed. Most writers agreed that flax was best sown broadcast (Fig. 71). Drilling flax seed was usually only advised where the crop was being grown for seed rather than fibre. Drilled flax required less than half the quantity of seed used in the broadcast method, but if sown in rows by drilling, there tended to be a relatively wide space to the side of each plant, which encouraged the development of short side stems whose fibre was of very limited value. One suggested way to prevent this was to sow half the seed by pulling the seed-drill along the field, and the other half by pulling it across the field. This could achieve a distribution which was described as ideal: 'If you were to draw the whole field out in square inches, there should be a seed at every corner'.[39] Even in this case, however, the seed drill was still not recommended, as it buried the seed too deeply in the ground.

Sowing flax-seed evenly using the broadcast method was even more difficult than sowing grain. Flax seed is small and light, and easily blown about by a breeze, so a calm day was recommended as the best for sowing.[40] An added difficulty for the sower was that the shiny seed tended to slip too easily from the hand. Because of this difficulty, Hyndman recommended a slightly different technique to that used in sowing grain:

> Now I suppose you have the seed in your apron, take an handful of it, and begin to sow your ridge at the side towards the left hand, and ending at the side towards the right, then step forward, and sow as before, always observing never to sow the handful at less than six or eight casts, throwing it a good distance from you; you are never to sow your seed with the broad-cast as wheat, oats . . . or the like is sown; but with your thumb straight forward.[41]

Several writers advised that flax, like all plants with small seeds, should be sown thinly at first, and the remaining seed then used to thicken up the distribution. Seed barrows were sometimes used for sowing, but the most common mechanical aid was a type of seed-fiddle, which may have been the prototype of those already described in the chapter on sowing grain (Fig. 42). The fiddle in question was the 'Little Wonder', patented by H. Adams of Garry, Ballymoney, County Antrim. An early twentieth-century description of this machine does not seem very different from those of grain-sowing fiddles.

> This machine, which is slung over the shoulder of the operator, consists of a hopper, from which the seed is delivered on to a disk, which is made to revolve by means of a string attached to a bow, hence the name "fiddle". The string is passed round a drum carrying the disk, and by moving the bow the disk is rotated and the seed thereby scattered. Whether sown by hand or machine, the land should be marked off in strips, each the width of a cast — 12ft. where the fiddle machine is used.[42]

A slightly different mechanism was used to revolve the disk on a sowing machine donated recently to the Ulster Folk and Transport Museum. On this machine, the disk was turned by a handle attached to one side.

Overall this sower is slightly smaller than the 'fiddle' type, and was claimed by the donor to be used only for sowing flax.[43]

Agriculturalists agreed that the depth at which flax-seed was planted was as crucial as the spacing between seeds. In 1774, Hyndman argued that flax-seed was best covered by trenching, and claimed that he was the first person in Europe to have discovered this, although we have already seen that grain-seed planted on ridges was often covered by shovelling earth from adjacent furrows. In Hyndman's method the flax-seed was sown broadcast. He argued that it did not matter if seeds fell in the furrows between ridges, since they would be put on top of the ridges by the shovelling. Enough mould was dug and shovelled from the furrows to cover the seed to a depth of one or two inches, depending on the nature of the land. Stones, clods and weeds were than raked off the tops of the ridges with a short toothed rake. Hyndman advised against rolling the newly sown ground, for several reasons. He argued that the feet of men and horses would push the seed out of place, that the roller would compress clods and stones into the ground, and that it would also break down the edges of the ridges.[44] He listed the general advantages of trenching flax as follows:

1. Hereby all the seed that is sown will grow.
2. There will be no after-growth.
3. All the flax will be nearer one length . . .
4. Birds can pick none of the seed; so the same quantity of seed will sow more land . . .
5. It will not be so apt to lodge, as in general its roots are deeper in the earth.
6. The furrow is a lane for the air . . .
7. The furrow is a drain in wet land, and even in dry land, in time of a great fall of rain, and does no harm to dry land in the driest season.
8. . . . if properly covered [the seed] will vegetage as soon in dry as in moist weather.
9. We may sow our seed earlier . . . for the trenches will carry off the great falls of rain or snow.[45]

Hyndman admitted that the usual objections to trenching could be raised; that the furrows wasted land, and that covering the seed required a lot of labour. He claimed, however, that in his method the furrows took up only about one twelfth of the surface area, and that the greater returns obtained from the crop would justify the extra labour.[46]

Hyndman criticised what he recognised as the common method of covering flax-seed, using a harrow, because he argued the seed was covered to different levels, sometimes being left on the surface and sometimes being buried as deep as four inches. The surface seed was exposed to birds, while the varying depths at which the surviving seed was covered led to the growth of plants of different lengths and thickness.[47] Despite these objections, however, harrowing and then rolling the newly sown ground remained the most common method of covering seed. Light seed-covering harrows were used, sometimes being pulled first along the field, and then across it. In the early nineteenth century, Mr Besnard, 'Inspector-General of the Linen and Hempen Manufacture for the Provinces of Leinster, Munster and Connaught', also recommended the use of 'a pretty large thorn bush, whereon you fix a weight of timber sufficient to make the thorns enter the ground'.[48] If it was not possible to roll the ground immediately after sowing, this could be postponed until the plants were 3-4 inches high.[49]

Care of the growing crop consisted mainly of weeding. If the ground had been well prepared, it was relatively weed-free, but despite Sproule's claim, mentioned above, that flax plants grew too closely together to allow grass to grow along with it, the Irish Department of Agriculture pointed out that the small leaves on growing flax plants meant that other plants, including weeds, could easily grow amongst the crop.[50] Most writers advised that weeding should begin when the flax plants were about four inches high. Between six and eight inches, the stalks were thought to have become too brittle to regain an upright position after having been flattened by weeders. During the eighteenth and early nineteenth centuries weeding was usually done by women and children who crawled over the ground on their hands and knees.[51] Near Armagh, Arthur Young was told that ten women could weed an acre of flax in a day.[52] To assist the young plants which had been flattened during weeding to become straight again, it was recommended that the weeders should work into the wind, which would then blow the shoots back into an upright position.[53] Apart from minimising the extent to which plants would be flattened, the main concern was that pulling out weeds would not disturb the roots of the flax. It was suggested that this could be avoided by weeding the crop just after rainfall which would have made the surface soil compact, and that larger weeds should be cut rather than pulled.[54]

Pulling and Rippling Flax

Flax grown for manufacturing linen was pulled by hand until the 1940s, as cutting the stalks wasted the good-quality fibres near the roots (Fig.

Fig. 72. Pulling flax, near Cushendun, Co. Antrim, c.1920 (WAG 1011).

72). After pulling, the only major processing operation carried out on the farm was steeping or 'retting' the flax.

Flax grown for fibre was usually ready for pulling about one hundred days, or fourteen weeks, after sowing.[55] The crop was pulled about three or four weeks after the first blossoms had appeared. Exactly when the crop was ready was a matter of debate, however. Flax grown for fine linen was pulled earliest, but if pulled too soon the fibres were soft and weak, and the yield was low. Hyndman also warned that a lot of the fibres would rot when put into water to steep.[56] It was also generally agreed, however, that if left too long, only coarse linen could be manufactured from the crop. Most writers advised that the plants should be pulled when the bottoms of their stems had begun to turn yellow and to lose the small leaves which grew up the entire length of the stalks. Agriculturalists differed as to the extent to which the shedding of leaves should be allowed to continue, some advising harvesting after the leaves had fallen off the bottom half of each stem, and others recommending that two thirds of the stem should be bare of leaves before pulling.[57] The development of the seeds inside the bolls at the top of the plant stalks could also be used as a test of readiness. It was agreed that the seed bolls should have begun to turn yellow, and that when cut open should have seeds which were 'no longer milky but firm, and somewhat brown in colour'.[58] It was also recommended, however, that flax which had been twisted or flattened by rain should be pulled as quickly as possible to minimise damage to the fibres. Also, in very bright, sunny weather the base of the plant stems sometimes turned rusty or reddish brown. If this happened, the resulting fibre could be discoloured and brittle, and again farmers were advised to pull the crop as quickly as possible to lessen the damage.[59] Agriculturalists also recommended that if some patches of the crop ripened before others, they should be pulled separately. Careful harvesting also involved pulling long and short, and coarse and fine, plants separately. If this was not done, many of the fine, short fibres would be lost in the later processes of breaking and scutching.[60]

Ridges or ploughed flats were used as guidelines to mark off patches to be pulled. Pulling a single handful of flax is relatively easy, but harvesting an entire crop was a backbreaking task. Each handful of flax was grasped about half way up the plant stems and pulled upwards and slightly to one side, and some skilled workers could use both hands for pulling. Four handfuls of flax made up a 'beet' or sheaf. The rate at which flax could be pulled obviously varied between workers, but one recent estimate suggests that one worker could harvest an acre in seven days.[61] In Ireland, beets were almost always bound using bands made from rushes. Bands could be made from flax, but this was considered wasteful as the fibres might be damaged. Beets were usually about eight inches in diameter, but careful farmers made them only four inches wide. This lessened the danger of heating inside the beet before it was steeped, and also meant that during steeping 'retting' was more likely to be even.[62].

It was generally agreed that a single flax crop could not produce both good seed and good fibre. Very little flax was grown for seed in Ireland, but the bolls were sometimes removed for animal fodder. The most common technique of removing seed was known as rippling (Fig. 73):

> As the flax is pulled, the separate handfuls which form a beet are crossed diagonally, so that when the beets come to the rippler the branched ends are not entangled, but each handful may be conveniently lifted without disturbing the remainder of the beet. The 'rippling comb', consisting of a set of iron teeth about 10 in. long, is fixed across the plank supported in some convenient way, 18 in. above the ground. The two workers sit astride the plank facing one another, one on each side of the comb, and alternately strike through the comb a handful of flax. By pulling the flax towards them the seed bolls are combed off, and these then fall on to a sheet spread beneath.[63]

The plank to which the rippling comb was attached was sometimes fixed

Fig. 73. Removing seed bolls from flax by rippling, near Toome, Co. Antrim, c.1920 (WAG 1024).

across the back of a cart, the seed falling into the body of the cart being very easily collected.

Seed intended for fodder rather than planting could be rippled from the flax within twenty-four hours of pulling, but if it was intended for sowing, the flax was dried before rippling, by first being spread and then put into small sheaves and stooks. If the weather was fine, the seed would become dry and ripen after about a fortnight. The flax could then either be rippled or stacked until the following spring. Once rippled, the bolls were best spread in an airy loft and turned frequently to prevent heating. When completely dry, the bolls were crushed by threshing or rolling, and the seed was then separated by winnowing.[64]

Some writers claimed that even where the seed was not to be used at all, rippling was essential to preserve the quality of the flax fibre: 'It is recommended to ripple off the bolls before watering. It is supposed they are in a great degree the means of discolouring the flax, and hastening putrefaction'.[65] However, early experiments by the Irish Department of Agriculture led their researchers to suggest that because of the difficulty of obtaining labour, and the uncertainty of the weather which made the flax harvest difficult anyway, rippling was not remunerative. They also added that the rippling process might damage the flax fibres.[66] Some Ulster farmers claimed that steeping flax with the seed bolls still intact was actually beneficial to the fibre: 'It puts a shine on it'.[67]

Retting Flax

In the retting process the woody central parts of flax plant stalks, or 'shous', were separated from the surrounding fibres. 'Dew' retting was sometimes practised, the flax being spread out over a grass or stubble field, and occasionally turned. Rain and dew provided the moisture which eventually separated the woody stalks from the fibres. This system was more common in England than in Ireland, however. Most Irish farmers retted their flax crops by steeping them in a dam or pond.

Flax which had been rippled and dried could be stacked before being retted during the following spring. In 1825, the North West of Ireland Society advised against this, flax in this state not being so easily watered, as well as being 'harsher and much inferior . . . [and] cut by vermin in the stack'.[68] At the end of the same century, however, it was claimed that 'some of the best flax fibre is obtained from flax that has stood twelve months before it was steeped'.[69] Whichever claim was true, most Irish flax was steeped very soon after pulling. In 1774, Hyndman criticised a practice which he claimed was common, of pulling flax too early, and then letting it stand in stooks or 'shocks' for anything up to twelve days. He claimed that beets stooked in this way would overheat and rot within thirty hours, a problem which 'the country suffers so much by'.[70] In 1908 it was claimed that the crop could be stooked, root ends up, for a few days without damage, if the weather was dull, but that in sunny weather it should be steeped within twenty-four hours.[71]

The water in which flax was steeped had to be carefully selected. Because its temperature was crucial, steeping was usually started before mid-August, and only very rarely after mid-September. The quality of the water was also important. Hyndman advised farmers to use lake water, or water let into a river or stream. A river could also be used, he claimed, but water standing in a clay pit was very unsuitable.[72] The *Munster farmer's gazette* also urged farmers to be

> very choice of the water; it must not be an airy swift current . . . it must not be hard water that partakes of mines or minerals . . . it ought not to be in bog waters, because they generally are stained by a shrub that grows on the bogs, called bogalders, which stain can seldom be got out by the best bleaching.[73]

Fig. 74. Putting flax into a dam to steep, c.1920 (WAG 1013).

Water containing iron or lime was also rejected as unsuitable.

Flax dams or ponds were usually dug near a stream or lake from which they could be easily filled. Dams tended to be long and narrow (Fig. 74). One fairly typical dam in Ulster was about seventy feet in length and seven feet in width. The narrowness allowed workers to fork the beets into the dam from both sides. It was not recommended that dams should be made deeper than four feet, as water deeper than three and a half feet was usually too cold at the bottom to allow easy retting.[74] Hyndman recommended that the dam should be lined with fern or rushes to lessen the muddiness of the water.[75]

The arrangement of beets in a dam varied. Early writers recommended that they should be set nearly upright, 'The top being the most difficult to water, ought to lie where the water is warmest'.[76] However, throughout the past two centuries, and in the present one, beets were often put down almost flat with the root ends only slightly lower than the tops. Two layers might be arranged in this way.[77] Careful farmers kept beets of coarse and fine flax separate in the dam, as the coarser plants retted more quickly and should be removed sooner. The beets were weighed down with stones or sods, the stones being between 14 and 28 lbs in weight, 'as big as your head'.[78] The degree of compression was important. If weighted down too tightly, the plants retted unevenly. On the other hand, the retting process involved

Fig. 75. Removing retted flax from a dam in Co. Louth, in 1783 (detail from a print by W. Hincks).

fermentation, and this often made the beets rise to the surface, after several days, so that more stones had to be added. In cold weather some farmers added cow manure to the water to encourage fermentation. The cows themselves could present problems, however: 'Cattle love flax water — it's like beer . . . You have to keep them away, they go mad for it!'[79] Some farmers preferred not to let any water flow into the dam during retting, but it was also argued that a slow trickle might be advantageous, since it was generally accepted that retting in a large volume of water produced fibres of a better colour than was produced from flax steeped in small dams.[80]

The length of time flax was kept in the dam varied between one and two weeks, depending on the type of flax, the quality of the water, and the temperature. Flax left too long in the dam became soft and tended to break during later processes in linen production. On the other hand, the woody stalks and fibres were difficult to separate if flax was under-retted. Some farmers believed that it was a worse mistake to under-ret than to over-ret. One test of the condition of retted flax was to 'break a few stalks in two places near the middle and about six inches apart; catch

Fig. 76. Spreading flax to dry, c.1920 (WAG 1015).

the broken bit of shou, and if it separates freely without breaking or tearing the fibre, it may be taken out'.[81] The beets could be removed using a drag fork, or by workers actually getting in to the dam and throwing them out (Fig. 75). This job had to be done carefully, as the flax was very tender just after steeping.[82]

Water in which flax has been steeped, despite its alleged attractions for cattle, is very poisonous to fish, and also smells extremely bad. Even in the early nineteenth century, farmers who steeped flax in small rivers or lakes were condemned for destroying fish.[83] It was sometimes claimed that flax water was good for land, but this was denied by later writers. Farmers were advised either to let the dam empty itself by seepage, or to let the water into a stream during flooding.[84]

Once beets had been removed from a dam, they were set upright in heaps to allow them to drain. After a few hours they were than carted to a fairly bare field, where the flax was spread in rows to dry (Fig. 76). The flax was turned several times during drying to prevent discolouring. Drying could be completed within six days in fine weather, but might take up to a fortnight. A farmer could tell that the plants had dried when a large proportion of the stalks formed a 'bow and string, caused by a contraction of the fibre from the woody stalk'.[85]

Some Ulster farmers adopted a system of drying which they described as similar to hay-making, and which was described as a Scottish method

Fig. 77. Flax being dried after retting. In the background, single sheaves or beets stand as 'gaitins'. The stooks in the foreground were made when the flax was almost dry, but before being stacked, c.1920 (WAG 1068).

in an early nineteenth-century text.[86] Here, the flax was 'gaited' when it was 'about three-quarters dry'. A 'gaitin' was made the same size as a beet, but tied nearer to the top. The bottom of the plant stems was then spread out to form a wide base on which the gaitin could stand (Fig. 77). Between two days and a week later the beets were made into ricks which could be up to thirty feet in length. The flax could be stored in the rick, or made into stacks until it was taken for bruising and scutching. However, although these latter processes were included as part of the dual farming-linen economy of many small Ulster farms in the late eighteenth, and early nineteenth centuries, they are related to textile production rather than crop cultivation.

Methods of flax cultivation changed less during the last 250 years than those used in the cultivation of any other crop dealt with in this book. Until the Second World War, mechanisation was minimal. Experimental flax pullers were developed in Europe during the 1920s, and during the 1940s flax-pulling machines were manufactured by James Mackie and Sons Ltd. of Belfast (Fig. 78), but coincidentally these appeared just before the period when flax cultivation completely disappeared in Ulster.

Fig. 78. A mechanical flax-puller, of a type manufactured during the 1940s (UFTM, L1585/11).

Agriculturalists did discuss flax-growing techniques, however, and the familiar complaint was made, that Irish farmers were irrationally reluctant to adopt new methods. 'The only obstacle to our farmers' enjoying the same advantages in the cultivation of flax as their continental neighbours, is their obstinate adherence to old customs'.[87] Irish flax cultivation was especially criticised when compared to the techniques of production of the crop found in Holland. Throughout the nineteenth century, various institutions organised visits for Irish farmers to the Low Countries to learn their techniques, or brought over instructors to teach what they saw as better methods.[88] Most criticisms, however, were made of the ways in which Irish flax was processed (especially by steeping and scutching) rather than of how it was grown. Slovenly processing was claimed to lead to a poor display at markets:

> Irish flax is so badly cleaned, and made up in so careless a manner as to quantity, and so fraudulent with respect to quality, that all confidence with the English and Scottish purchasers has been destroyed . . . When brought to market, it is full of shoves, from the reedy part not having been properly

removed . . . The practice also of mixing long and short flax of different colours, in the same parcel, leads to numerous frauds.[89]

In discussing Irish methods of growing flax, agriculturalists were not nearly so severe. Arthur Young, for example, described flax fields at Clonleigh, County Donegal, as 'nothing but weeds and rubbish of all kinds', but added that 'the crop itself had the appearance of being good'.[90] In the mid-nineteenth century, Martin Doyle claimed that the north of Ireland could be classed along with the Netherlands as a part of Europe where flax cultivation was best understood.[91] Some Irish farmers might not have practised the careful preparation of soil, or the extended rotations necessary for good flax crops, but there was very little that agriculturalists could condemn wholesale. In 1901, the Irish Department of Agriculture concluded that only very detailed experiments would reveal ways in which flax cultivation might be improved: 'Enquiries in this Country and on the Continent disclosed the fact that comparatively little reliable information was available, and it became evident that if accurate information was to be obtained it would have to be furnished by a series of experiments conducted by the Department themselves.'[92]

The most significant influences on Ulster farming during the last two hundred years did not come directly from flax cultivation, but from the related linen industry. Eighteenth-century commentators, in particular, condemned the effects of the dual economy of weaving and farming on the Ulster countryside:

> View the North of Ireland; you there behold a whole province peopled by weavers; it is they who cultivate, or they beggar the soil, as well as work the looms; agriculture is there in ruins; it is cut up by the root; extirpated; annihilated; the whole region is the disgrace of the kingdom; all the crops you see are contemptible; are nothing but filth and weeds. No other part of Ireland can exhibit the soil in such a state of poverty and desolation . . . the cause of all these evils . . . [is] the [linen] fabric spreading over all the country, instead of being confined to towns.[93]

James McCully, writing a few years later, commented: 'To a stranger it must certainly appear paradoxical when he is told that the establishment of a prosperous manufacture, had almost ruined agriculture in the country where it is carried on'.[94] The paradoxical relationship between a flourishing linen industry and unsatisfactory farming practice can be paralleled in flax cultivation. Flax, closely associated with 'commercialisation' and 'modernisation', was one of the last crops to be mechanised, and was often grown on patches of land too small, or too steeply sloping, to be efficiently used in mechanised crop cultivation.

More than any other crop, flax illustrates one of the central arguments of this book: that no easy connection can be made between technological change and the advancement of farming.

NOTES

1. Young, Arthur, *A tour in Ireland*, vol. 1 (Dublin, 1780), pp.59 and 141.

2. Sproule, John, *A treatise on agriculture* (Dublin, 1839), p.358.

3. Clark, A.L., 'Historical sketch of the flax-growing industry', The Department of Agriculture and Technical Instruction for Ireland, *Journal*, vol. 3 (Dublin: H.M.S.O., 1903), p.688; Berry, Henry F., *A history of the Royal Dublin Society* (London, 1915), p.57.

4. Clark, A.L., *op. cit.*, pp.693 and 696.

5. *Ibid.*, p.702.

6. *Ibid.*, pp.689, 694 and 695.

7. Sproule, John, *op. cit.*, p.358.

8. Wright, R.P. (ed.), *The standard cyclopedia of modern agriculture and rural economy*, vol. 6 (London, 1908), p.12.

9. Crawford, W.H., *Economy and society in eighteenth century Ulster*, Ph.D. thesis (Belfast: Queen's University, 1982), p.82.

10. Clark, A.L., *op. cit.*, p.689.

11. *Ibid.*, pp.697 and 699.

12. *Ibid.*, pp.699-700.

13. Crawford, W.H., *op. cit.*, pp.82-103.

14. Hyndman, C., *A new method of raising flax* (Belfast, 1774), p.33.

15. Crawford, W.H., *op. cit.*, p.91.

16. Young, Arthur, *op. cit.*, vol. 1, pp.174-175.

17. Crawford, W.H., *op. cit.*, p.48.

18. Sproule, John, *op.cit.*, pp.358-359.

19. Young, Arthur, *op. cit.*, vol. 2, 'General observations', p.155.

20. Irish National Schools, *Introduction to practical farming*, vol. 1 (Dublin, 1898), p.54; Wright, R.P., *op.cit.*, vol. 6, p.12.

21. Young, Arthur, *op.cit.*, vol. 2, 'General observations', p.156.

22. Baldwin, Thomas, *op. cit.*, p.55.

23. Interview with Richard, Eugene, and Felix McConville, Dromore, county Down (Ulster Folk and Transport Museum tape R82.95).

24. Hyndman, C., *op. cit.*, pp.11-12.

25. North West of Ireland Society *Magazine*, vol. 1 (Derry, 1823), p.23.

26. Young, Arthur, *op. cit.*, vol. 1, pp.177-178.

27. North West of Ireland Society, *op. cit.*, vol. 1, p.31.

28. Hyndman, C., *op. cit.*, pp.13-14.

29. *Ibid.*

30. Hyndman, C., *op. cit.*, p.10.

31. Ulster Folk and Transport Museum tape R.82.95.

32. Young, Arthur, *op. cit.*, vol. 1, p.163.

33. *Ibid.*, p.183.

34. *Ibid.*, p.163.

35. Clark, A.L., 'Flax seed for sowing purposes', The Department of Agriculture and Technical Instruction for Ireland, *Journal*, vol. 4 (Dublin: H.M.S.O., 1904), p.269.

36. Wright, R.P., *op. cit.*, vol. 6, p.13.

37. Young, Arthur, *op. cit.*, vol. 1, p.163.

38. Sproule, John, *op. cit.*, p.362.

39. Ulster Folk and Transport Museum tape R82.95.

40. North West of Ireland Society, *op. cit.*, vol. 1, p.31.

41. Hyndman, C., *op. cit.*, p.16.

42. Wright, R.P., *op. cit.*, vol. 6, p.13.

43. Ulster Folk and Transport Museum, Specimen number 209.1983.

44. Hyndman, C., *op. cit.*, p.17.

45. *Ibid.*, p.20.

46. *Ibid.*, p.21.

47. *Ibid.*, pp.18-19.

48. *The Munster farmer's magazine*, vol. 5 (Cork, 1817), p.148.

49. Wright, R.P., *op. cit.*, vol. 6, p.13.

50. Department of Agriculture and Technical Instruction for Ireland, 'Flax experiments, 1901', *Journal*, vol. 2 (Dublin: H.M.S.O., 1902), p.640.

51. Sproule, John, *op. cit.*, p.364.

52. Young, Arthur, *op. cit.*, vol. 1, p.165.

53. Baldwin, Thomas, *op. cit.*, p.56.

54. Wright, R.P., *op. cit.*, vol. 6, p.13; *The Munster farmer's magazine*, vol. 5 (Cork, 1817), p.148.

55. Wright, R.P., *op. cit.*, vol. 6, p.14.

56. Hyndman, C., *op. cit.*, p.23.

57. *Ibid.*, p.22; North West of Ireland Society, *op. cit.*, vol. 1, p.38.

58. Wright, R.P., *op. cit.*, vol. 6, p.13.

59. *Ibid.*

60. North West of Ireland Society, *op. cit.*, vol. 1, p.38.

61. Ulster Folk and Transport Museum tape R82.95.

62. Wright, R.P., *op. cit.*, vol. 6, p.14; Ulster Folk and Transport Museum tape R82.95.

63. Wright, R.P., *op. cit.*, vol. 6, p.14.

64. North West of Ireland Society, *op. cit.*, vol. 3, p.104; Wright, R.P., *op. cit.*, vol. 6, p.14.

65. North West of Ireland Society, *op. cit.*, vol. 1, p.39.

66. Silcock, James *et al*, 'The influence of rippling', The Department of Agriculture and Technical Instruction for Ireland, *Journal*, vol. 3 (Dublin: H.M.S.O., 1903), p.680.

67. Ulster Folk and Transport Museum tape R82.95.

68. North West of Ireland Society, *op. cit.*, vol. 1, p.39.

69. Baldwin, Thomas, *op. cit.*, p.57.

70. Hyndman, C., *op. cit.*, pp. 23-24.

71. Wright, R.P., *op. cit.*, vol. 6, p.14.

72. Hyndman, C., *op. cit.*, p.25.

73. North West of Ireland Society, *op. cit.*, vol. 1, p.39; Wright, R.P., *op. cit.*, vol. 6, p.14.

74. North West of Ireland Society, *op. cit.*, vol. 1, p.39.

75. Hyndman, C., *op. cit.*, p.26.

76. North West of Ireland Society, *op. cit.*, vol. 1, p.40.

77. Wright, R.P., *op. cit.*, vol. 6, p.14.

78. Ulster Folk and Transport Museum tape R82.95.

79. *Ibid.*

80. Wright, R.P., *op. cit.*, vol. 6, p.14.

81. Baldwin, Thomas, *op. cit.*, p.49.

82. North West of Ireland Society, *op. cit.*, vol. 1, p.45.

83. The Munster farmer's magazine, op. cit., *vol.* 4, p.255.

84. Wright, R.P., *op. cit.*, vol. 6, p.14.

85. Baldwin, Thomas, *op. cit.*, p.50.

86. Ulster Folk and Transport Museum tape R82.95; the North West of Ireland Society, *op. cit.*, vol. 1, p.45.

87. North West of Ireland Society, *op. cit.*, vol. 2, p.194.

88. Doyle, Martin, *A cyclopaedia of husbandry*, rev. ed. W. Rham (London, 1844), p.250; Clark, A.L., *op. cit.*, (1903), pp.690-691; Department of Agriculture and Technical Instruction for Ireland, *op. cit.*, vol. 2, p.636.

89. North West of Ireland Society, *op. cit.*, vol. 2, pp.195-196.

90. Young, Arthur, *op. cit.*, vol. 1, p.238.

91. Doyle, Martin, *op. cit.*, p.239.

92. Department of Agriculture and Technical Instruction for Ireland, *op. cit.*, vol. 2, p.636.

93. Young, Arthur, *op. cit.*, vol. 2, 'General observations', pp.162-163.

94. McCully, James, *Letters by a Farmer* (Belfast, 1787), p.7.

CHAPTER 9

The Grain Harvest

Oats, wheat and barley have been the main cereal crops in Ireland for hundreds of years. By the end of the seventeenth century exports of grain and flour to England, Scotland, and even France, were well established[1] and this trade continued throughout the next two centuries, rising and falling with the relative yields of harvests in Ireland and abroad. There were periods when Ireland was a net importer of grain, but these were short-term until the end of the nineteenth century.[2] Grain cultivation was encouraged by state institutions. In 1758, for example, a bounty was granted on the inland carriage of grain and flour to the Dublin market, while Foster's Corn Law of 1784 gave bounties on grain exports. The Napoleonic Wars, 1793-1815, were a boom time for Irish farmers, and although a severe slump followed the wars, exports of grain and flour doubled during the 1820s, and continued to rise steeply throughout the 1830s.[3] The scale of exports during the 1820s led to the claim that Ireland had become 'the granary of Britain'.[4] Home consumption of grain was even greater than the amounts exported, and even during the 1820s, two thirds of the grain grown was used in Ireland.[5] In the second half of the nineteenth century, however, there was an almost continuous decline in cereal cultivation. The table below summarises this decline, but also shows the relative importance of each of the major grain crops in Ireland:[6]

Acreage of Crops, 1855-1901

Year	Oats	Barley	Wheat
1855	2,118,858	445,775	226,629
1860	1,966,304	466,415	181,099
1870	1,650,309	259,849	241,285
1880	1,381,928	148,708	218,016
1890	1,221,013	92,341	182,058
1901	1,099,335	42,934	161,538
Percentage Decrease	48.1%	90.4%	28.7%

Oats were by far the most important Irish cereal. Oatmeal was a common element of diet, especially in the north, and was also widely used as animal bedding, and in the eighteenth century was often used as a substitute for hay. Agriculturalists agreed that the soils and climate of Ireland were particularly suited to the cultivation of the cereal. Several commentators remarked that this was clearly shown by the fine crops produced, despite the overcropping so frequently observed and condemned throughout the eighteenth and early nineteenth centuries.[7] Oats presented fewer problems in harvesting than the other major cereals. The light straw was easily cut, although the standing crop was at risk from high winds or heavy rain which could flatten or twist the grain. Another problem shared with wheat and barley was that grain was lost if the crop was shaken during cutting. In Ireland this problem was accentuated by the relatively late harvest. Compared to much of the rest of Europe, Irish summers are cool and moist, which means that grain ripens more slowly. In 1835, it was claimed that while oats and barley were usually cut in England during July, in Ireland they were not cut until August or even September.[8] The longer a cereal crop stands uncut, the looser the grain becomes in the seed heads. It was estimated by one late nineteenth-century writer that the quantity of both wheat and oats lost in Ireland during the harvest by 'shelling and shedding' nearly equalled the amount of seed sown.[9] One unusual solution to the problem was suggested by a Belfast newspaper in 1750:

> As the corn harvest is approaching, it is recommended to the farmer, when high winds happen and he is afraid of his corn shaking, that he immediately break the stalks off all his ripe corn about the middle of the stalk; which may be done in a very short time with the hand: this will prevent the high winds doing much, if any, damage. None but the ripe corn will readily shake. This method has been pursued with good success in some parts of the country, but hath not hitherto generally prevail'd. A poor man having one ridge of rank corn, and no hands to cut it down, when a violent wind arose; he roll'd himself over that ridge, and then cut it down at leisure, and thereby preserved it from a great shake, which did considerable damage to his neighbour's grain. But had he only broke the stalks, as above directed, it would have been much easier afterwards to cut down.[10]

Despite the newspaper's approval, however, no evidence has been found that either technique suggested became widely used! A more conventional solution to the problem, suggested by several agriculturalists, was that the grain should be cut three or four days before it was dead ripe. It was claimed, however, that even this advice was not followed by many Irish farmers.

Wheat was generally accepted as being less suited to Irish conditions than oats, wheat growing best in warm, dry conditions, and on heavy clay loams. In the late eighteenth century Arthur Young found that some farmers in county Armagh believed that wheat would not grow there at all. This was disproved by the Anglican primate who produced a 'very fine and very clean' experimental crop.[11] Most wheat cultivation was, however, confined to the sunnier, drier south-eastern counties. The grain was used most of all for bread, but the straw was also valued. Wheat straw is harder and stronger than oat straw. This limited its use for animal fodder, but meant that it was good for thatching, and for the construction of a wide range of objects made in Ireland from plaited or twisted straw. One nineteenth-century variety, Whittington's, could grow to over six feet in height.[12] Where wheat was grown successfully, as in county Kilkenny, wheat crops enabled farmers 'not only to pay their rent and subsist without difficulty, but by the increasing value of corn to acquire additional profits.'[13] Autumn-sown 'winter' wheat was recommended as more suitable to Ireland that 'spring' wheat. However, in order that the crop could survive heavy winter rains without becoming waterlogged, it had to be cultivated on higher, narrower ridges than other cereals. This, combined with the relative heaviness of the ripened grain heads, gave rise to special problems during harvest.

The cereal crop presenting most difficulties in harvesting however, was barley. Barley was used in brewing beer, ale and porter, and in the distillation of poteen and whiskey. However, deciding on the right time to harvest the crop required careful consideration. If it was cut early, the grain suffered: 'It not only shrivels much, but, when it is reaped with any green on the grains, it assumes a bleached white colour instead of the rich golden-yellow colour which it exhibits when not discoloured by rain'.[14] Grain which was not dead ripe also had to be left longer in the field, during which time it was very susceptible to overheating, if bound or stacked.[15] If, on the other hand, the crop was left standing until it was fully ripened, the stalks became brittle and the heads were liable to break off during cutting.[16] Over-heating remained a problem here also, since the crop tended to be weedy, and fermentation could occur if the grain was stacked, or even stooked, before the weeds had withered.[17]

Apart from oats, wheat, and barley, rye was the only cereal cultivated in significant quantities in Ireland. The grain was hardly used at all for bread, but the straw was valued for thatching. In the late eighteenth century, rye had some importance as the first crop grown on reclaimed land by large-scale improvers.[18] Rye presented relatively few problems in harvesting, and was the only major cereal which agriculturalists advised could be left standing until the crop was dead ripe. Like flax, rye was sometimes pulled rather than cut.

Until the late nineteenth century, almost all crops were harvested by hand in Ireland, and of the available implements, sickles and hooks were by far the most widely used. Scythes were common by 1800, but were mostly kept for mowing hay. It was not until the late nineteenth century that they became widely used for harvesting grain.

Sickles and Hooks

Hand tools for harvesting grain vary widely throughut the world, and classifying particular implements as sickles, hooks or scythes can prove difficult.[19] Fortunately, during the last two centuries differences in the size and shape of the implements used in Ireland make the classification fairly easy. The scythes used have been much larger than either hooks or sickles, and their blades less curved. The most important difference between reaping hooks and sickles has been that the former have smooth-edged blades, the latter have blades with serrated, saw-like edges.

Artifacts identified as reaping hooks have been found on Irish archaeological sites ranging in date from the Bronze Age to the medieval period.[20] Literary evidence for the use of reaping hooks is also ancient, references being found in the earliest texts. The terms used in these are *serr* and *corrán*[21] (The latter word is still used in modern Irish.) Surviving specimens suggest that until the medieval period, smooth-bladed hooks were the dominant harvest tool, but from late medieval times, literary references to toothed sickles (Irish: *corrán círe*) became increasingly common. By the seventeenth century the sickle had become one of the distinctive emblems of the Irish peasant[22] (Fig. 79). The dominance of toothed sickles continued well into the nineteenth century, even after the earlier blacksmith-made tools began to be replaced by imported foundry-made implements. These later, mass-produced, sickles often had the teeth on their blades tipped with steel and the blade curved to form a semi-ellipse, elbowed just above the handle, this being generally accepted as the most efficient cutting shape. By the end of the nineteenth century, foundry-made sickles were used throughout Ireland. One specimen in the National Museum in Dublin (Reg. No. 1967.93), for example, was used on Inishbofin island, county Galway. Agricultural implement makers, like Tyzack's of Sheffield, produced many patterns

Fig. 79. A young woman holding a toothed sickle. This illustration accompanied a story, 'Handsome Kate Kavanagh', which was set in the Barony of Forth, Co. Wexford (*Irish penny journal* (6/2/1841), vol. 1).

of sickles and reaping hooks. Two sickles advertised by the firm in the 1923-24 catalogue are described as 'Irish.'

The main advantage sickles had over smooth-bladed reaping hooks was that they did not require frequent sharpening. Hooks could cut grain faster than sickles, however. Some evidence from Scotland suggests that, although sickles were the tool most favoured by Irish reapers, hooks may also have been fairly common. It has been claimed that since the eighteenth century, Irish migrant harvest workers in Scotland favoured smooth-bladed hooks,[23] and in 1831 the agriculturalist J.C. Loudon claimed that reaping hooks had actually been introduced into western and south-western Scotland by Irish harvesters.[24]

Several techniques can be used in reaping grain with sickles or hooks. A method common to both implements involves taking a handful of straw in one hand and cutting it with the blade held in the other (Fig. 80). Reaping using this technique can be either 'high', when only the ears and a short part of the straw are cut, or 'low', when the straw is cut

Fig. 80. Paddy Storey reaping oats with a toothed sickle near Toome, Co. Antrim, c.1920 (WAG 298).

as near to the ground as possible. These were generally agreed to be the neatest ways of cutting grain, since the reaper, by taking hold of the stalks, just below the seed heads, could leave behind many of the weeds growing lower down. Another faster method of reaping was also possible with smooth-bladed hooks, however. This involved hacking or slashing the standing grain, which might be held in position with a small fork made from furze. In England this technique was known as 'bagging' or 'fagging', and large 'bagging hooks' were developed for it. These larger hooks were known in south-eastern Ireland, but in general Irish reaping hooks were smaller (Fig. 81).

In a large field, reapers were followed by binders who tied sheaves, each made up of about three handfuls of grain. In Ireland, as in Scotland, reapers generally worked along the ridges on which the grain was grown. A common arrangement of reapers and binders in a large field was described by Martin Doyle:

> The mode of reaping differs in Great Britain and Ireland. The British labourer first cuts a small handful for the band, which he deposits and then (cutting across the ridge) he lays across the band as many handfuls as will constitute a sheaf, which he himself generally binds. The Irish reaper, being a more social animal, has three or four companions on the ridge,

Fig. 81. Mary and Ellen McCaughey of Corleaghan, Co. Tyrone, reaping and binding oats cut with a smooth-bladed reaping hook, c.1910 (UFTM, L918/9).

according to its breadth; one cutting a passage along the furrow, and ledging the several handfuls across it, and the others reaping a breadth of about eighteen inches each, every pair of reapers ledging together, but none of them preparing bands for their fair followers . . . In Scotland binders are generally provided, which is a great saving of time for the reaper, but in a much less proportion than in Ireland, where one binder (female) is allowed for two men.[25]

In fact, other accounts suggest that one binder had sometimes to work for more than two reapers, but Doyle's description makes the division of tasks between reapers and binders clear.

The arrangement of reapers across a field is interestingly described in the preface to a late nineteenth-century Ulster poem:

The leader of the 'boon', or band is 'stubble-hook', so called from his being employed on the open plot next to those which have been shorn: while 'cornland' occupies the ridge next to the standing grain, and may be looked upon as the driver. The shrewd farmer generally chooses two of his best shearers for these situations. He knows that each reaper from the leader to the driver is supposed to keep about the 'making' of a sheaf in the rear of the hook immediately preceding him: that the line thus formed is, under ordinary circumstances, to be kept unbroken; and that, therefore on the 'stubble-hook' and 'corn-land' depend, in great measure, the amount of labour to be accomplished by the hooks at work between them.[26]

It has been argued, on the basis of Scottish evidence, that women workers were valued for reaping with sickles, their bodies being well adapted physiologically to the stooped position required of the reaper.[27] Irish evidence, however, tends to follow a broader pattern identified throughout Europe since the medieval period.[28] Both men and women reaped with sickles and hooks, but scythes were almost entirely used by men.

Scythes

Literary evidence, from Gaelic sources, for the use of scythes in Ireland is relatively recent, clear references only appearing from the late sixteenth century. It has been suggested that the term used for scythe (*spel*) may have been a loan word taken from Middle English.[29] Scythes were, however, used on Norman manorial estates in Ireland. Labour services required of a *betagh* included the mowing of hay with a scythe. This early evidence is largely confined to Leinster.[30] However, as we have seen in an earlier chapter, by the eighteenth century, references to the use of scythes in hay-making had become much more common. The widespread use of scythes is indirectly suggested by Arthur Young's erroneous explanation of the meaning of the term 'spalpeen' (Irish: *spailpín*), which was applied to temporary migrant labourers. Young claimed that the term was derived from two words: 'Spal in Irish is a scythe a peen a penny, that is a mower for a penny a day'.[31] In fact the term is probably derived from the Irish *spailp*, meaning a spell or a period, and thus referring to the temporary nature of the spalpeen's employment. Young's mistake, however, does suggest that either he, or his informant, regarded the use of scythes as a typical task carried out by these workers.

Import records show that by the late eighteenth century, large numbers of scythes were being brought into Ireland from Britain. By 1800, the use of scythes in the grain harvest was beginning. In county Kilkenny, Tighe found that 'a few farmers have occasionally mown their spring corn: mowers do not like to undertake it, and say it spoils their scythes: it is constantly practised by Mr. Robert St. George, who advises

G

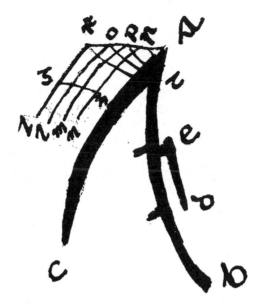

Fig. 82. A cradle-scythe drawn by Amhlaiobh ó Súileabháin in 1831: a,b: crann na speile d,e: na dornóga a,c: an speal a,f,i,n: an cliabhán fi,ol,pm,rn: plata an cliabháin i gcuspa don speal (Ó Súileabháin (1930, 1936), p.66).

that any kind of crop, except wheat should be mown with a cradle scythe'.[32]

A story originating in the late nineteenth century, and recorded in the Mourne mountains, county Down, by the folklore collector Michael J. Murphy, emphasises the relatively late use of scythes for harvesting grain. A migrant labourer who had been doing harvest work in Britain brought a scythe back with him. He demonstrated the speed with which grain could be cut using the implement to a group of increasingly impressed neighbours.[33] This story suggests the important point that many small Irish farmers may in fact have learned the technique of harvesting grain using scythes while working in England and Scotland. (It also has been claimed, conversely, that many Irish harvesters in nineteenth-century Britain would not, or could not, use scythes, and this slowed the introduction of scythes as grain cutting implements in some areas.[34])

Scythes with a 'cradle' attachment behind the blade to catch the newly cut swathe seem to have been first used in south-eastern Ireland. The diarist, Amhlaiobh ó Súileabháin, made a diagram of a cradle scythe (Irish: *cliabhspeal*) in 1831 (Fig. 82). His reaction to the implement, which he saw in use near Kilkenny, suggests the limited extent to which cradles were used:

Fig. 83. Mowing corn with scythes near Toome, Co. Antrim, c.1920 (WAG 1157).

> I saw a new arrangement on a scythe today, cutting oats for George Clinton. There was a cradle or cage mounted on the scythe . . . This scythe brings the [whole] produce of the ground with it, both straw and grain, for it cuts low, and lays it straight in the swathe . . . A mower with a cage scythe would cut as much oats or barley with this apparatus as four sickle men.[35]

Later in the nineteenth century the use of cradle attachments became more widespread. One specimen in the National Museum in Dublin (Reg. No. 1932:130), for example, came from Tuam, county Galway. An Irish National school-book, published in 1868, suggested that farmers could make their own cradles: 'A skilful man can mow corn very well with a common scythe . . . a few pieces of strong wire should be fixed to the sned [shaft] to catch the crop as it falls on to the scythe, and assist in getting it into a swathe'.[36] Other simple cradles were made by bending a twig to form a bow behind the scythe blade, or by stretching sacking or canvas over a curved piece of wire.

The most common scythe used in Ireland before the twentieth century had a long straight shaft or sned (Fig. 83). Two handles were fixed to the shaft at right angles to one another, and the distance between these was set for individual mowers. The angle of the scythe blade to the sned was also adjusted for different crops, using a piece of wire known as the 'grass-nail' which linked the blade and sned. Improving landlords,

Fig. 84. A photograph from Co. Tyrone, showing two girls making a stook from four sheaves of oats, tied together at the top, while in the background an older man is sharpening a scythe with a strickle, c.1910 (UFTM, L918/8).

returning migrant labourers and agricultural implement salesmen were probably responsible for the introduction of the wide variety of scythe types used in Ireland in recent times. Scottish Y-shaped shafts of wood and metal, and English and American S-shaped shafts were widely known.

Dineen's *Irish-English dictionary* makes an interesting distinction between the *speal Ghaedhealach* (Gaelic scythe) and the *speal Ghallda* (modern scythe).[37] The Gaelic scythe is described as having a rivetless blade. This suggests that the distinction was between local blacksmith-made scythe blades and factory-made implements. However, some of the latter were also rivetless. The 1921 catalogue of Tyzack's Ltd. of Sheffield illustrates the latter type, describing them as 'solid-back' scythes.[38]

Scythes require frequent sharpening when in use. Scythemen nowadays recommend that this should be done every ten minutes. Both sharpening boards, or strickles, and scythe stones were widely used in Ireland (Fig. 84). The making of scythestones was well established by the late eighteenth century in areas where suitable stones could be quarried. Using a scythe to cut grain is a skilled task. Mowers generally cut along the edge of the crop, some swinging the blade towards the main body of standing grain, others away from it. As it approaches the grain to be cut, the 'heel' (the joint where blade and shaft meet) should be travelling along the ground. Mowing is, however, generally agreed to be much quicker then reaping with hooks and sickles. E.J.T. Collins has produced the following figures, based on English evidence:

Hand-tool	Cutting rate per day (in acres)	Worker days per acre (including binding and stooking)
Sickle (reaping low)	0.25	4.8
Reap-hook (reaping low)	0.33	4.0
Scythe	1.15	2.4[39]

As these figures show, however, the increase in cutting speed did not automatically lead to a proportionate saving of labour when scythes were adopted. Three or four people might be engaged in the scything operation, if the advantages of speed were to be used to the full. The scythesman might be followed by a band-tier, who tied the sheaves, and a stooker. The technique of 'rodding' or 'poling' required another worker, usually a young boy or girl, to walk alongside the scythesman with a pole or a wooden rake, with which they pressed the grain to be cut, against the standing crop (Fig. 85). Like the forks made of furze, used to hold grain being 'bagged' with a hook, this allowed easier cutting.

The late introduction of the scythe to the Irish grain harvest can be explained by the limited extent to which it could be efficiently related to other aspects of the common systems of cereal cultivation. It has been argued in earlier chapters that these systems were developed within the very limited resources available to most Irish farmers, and also within the limits imposed by the mild, damp Irish climate.

The high, narrow cultivation ridges which persisted in Ireland well into the nineteenth century were recognised by Amhlaiobh ó Suileabháin as a major obstacle to the use of the scythe, which required ground that was 'free from hills and hollows'.[40] We have already seen that by 1800 some 'gentlemen' farmers in county Kilkenny were experimenting with sowing wheat on almost flat land, but this was not judged to be successful.[41] It was not until after 1850, when large-scale

Fig. 85. 'Poling' oats with a hay-fork to make it easier for the scythe to cut the crop. This photograph was taken near Slieve Gullion, Co. Armagh, 1930s (photograph by Caiomhín Ó Danachair).

drainage schemes became widespread and cultivation ridges began to be flattened, that scythes could be effectively used.

The dominance of broadcast sowing techniques in Ireland also lessened the initial attractiveness of scythes. Grain crops sown in long straight rows by seed drilling machines could be weeded using hoes, but this was extremely difficult when seed was simply scattered on the ground. Weedier crops were best reaped with sickles or hooks, since as mentioned earlier, harvesters using these tools could to some extent select the grain for cutting, but leave the weeds behind. Scythesmen could not reap selectively, and this meant that farmers with sufficient labour could expect cleaner crops if they were cut using the smaller implements.

The lateness of the Irish harvest also hindered the introduction of scythes. The longer grain stands uncut, the more likely it is to be flattened or twisted by rain. Where this has happened it is extremely difficult to use a scythe. (The practice of cultivating grain on ridges partially solved this problem, since the standing grain on a ridge provided some support for the crop on adjacent ridges. However, as just pointed out, this solution itself precluded the use of scythes.) In 1835, the agriculturalist Burroughs also pointed out that grain left standing in a cool, moist climate developed thicker, harder straw. For this reason, he argued, Irish barley, for example, had straw which was too thick to be easily cut with a scythe.[42] This claim seems to be supported by some European evidence. Lighter, spring-sown grain crops, particularly oats, were the ones on which scythes were earliest used.[43]

Many agriculturalists since the eighteenth century have debated whether mowing with scythes shakes more grain out of the crop than reaping with hooks or sickles. Some Irish farmers certainly claimed this. As late as 1863, a correspondent in the *Irish farmer's gazette* claimed that in an experiment testing scythes against 'fagging hooks' on an area of twenty acres of wheat, the hooks had proved cheaper, and had knocked out less than half the wheat lost by scything.[44] Similar claims were also commonly recorded in eighteenth-century England and Scotland. Some observers, however, have argued that the reverse was the case, and that scythes actually shook grain less than sickles or hooks.[45] In considering the question, it is important to remember that, as Martin Doyle pointed out:

> All modes of reaping corn are attended with the danger of shaking out the grains. There are reapers with the sickle who have a provoking habit of switching about every handful of corn which they cut, and scattering around them the grains in profusion. . . Some mowers with a scythe [on the other hand] sweep around the swathe with a jerk which whips out the grain.[46]

Differences between agriculturalists may have arisen from the fact that their conclusions were based on the observation of different techniques of reaping with sickles or hooks. Despite the evidence of the experiment mentioned above, it seems likely that where the slashing or 'fagging' technique of reaping with a hook was employed, the grain may have been badly shaken. The technique where grain stalks were held in one hand while being cut with the other would almost certainly have shaken the grain much less. In those parts of Ireland where this technique was employed, it seems reasonable to conclude that less grain would have been lost than if the crop had been cut with a scythe.

It can be argued in general, therefore, that scythes would only have become attractive to most Irish farmers as grain-harvesting implements when drainage schemes allowed the old cultivation ridges to be flattened, systematic weed control was developed, and harvesting was begun earlier. The main advantage of using scythes — an increase in cutting speed — only became significant when farming was large-scale. Even on middle-sized Irish farms, labour needs were usually met by members of the family, supplemented by help from neighbours or hired workers. Many farmers, therefore, preferred to continue using the neater reaping techniques possible with sickles or hooks. This provides a clear example of an apparent improvement in technology which was not, however, always the method which the Irish small farmer would rationally choose. It is interesting, also, that even as conditions were changing to allow the efficient use of scythes, there were already being technologically superseded by the development of reaping machines.

Reaping Machines and Reaper-Binders

Ancient evidence for the use of reaping machines in Europe can be found in the writings of Pliny and Palladius, who described their use in the south of France.[47] but systematic attempts to develop horse-drawn reapers only began in England and Scotland during the late eighteenth century. The patent for such a machine was granted in 1799,[48] and this was followed by a spate of inventions, culminating in the much-praised reaper developed by the Scot, Patrick Bell, in the late 1820s. There is at least one piece of evidence that these English and Scottish experiments were paralleled in Ireland. In 1806, the *Belfast commercial chronicle* published a letter describing the demonstration of a machine at Moira, county Down, which the correspondent hailed as a success:

> To the Farming Society of the county of Down
> Gentlemen — I was so very fortunate a few days since, as to be present at the Rev. Mr. MacMullan's, Curate of Moira, where a new invented machine for cutting down standing grain was exhibited; as experiment, the constructor cut, in a field of oats, against seven of the best reapers in the field; the result was, on his ridge was double the quantity of sheafs as on theirs, and one over — his stubble was more level and two inches shorter than theirs; in fact, the unanimous opinion of at least sixty persons who were present (most of whom were farmers) was, that the machine effects the work of fifteen reapers, and its execution superior. Too much praise, I think, cannot be given to Mr Jellett, both for the invention and his condescension in exhibiting this machine where requested, as also his willingness to instruct others in its use, and to have similar ones made for any person wishing to have them, under his direction, by a mechanic in Moira, who he means to instruct for general accommodation . . . A FARMER.[49]

Unfortunately, like many contemporary reaper designs, Mr. Jellett's invention seems to have disappeared without trace.

The obscure early experiments in reaping machine design can be sharply contrasted with the excitement which was aroused by the appearance of two American reaping machines, McCormick's and Hussey's, at the Great Exhibition in London in 1851. Trials were held throughout Britain during the following years, and orders for the machines outstripped supplies. Similar excitement greeted the appearance of the machines in Ireland, although several commentators concluded that the earlier Bell reaper performed better at trials. A Hussey-style reaper, designed by a Belfast man, Richard Robinson, was shown at trials held in Belfast in 1852, but lost to a machine of Bell's design.[50] In 1853, Hussey and Bell machines, made by Crossley of Beverley, Yorkshire, were exhibited at the Irish Industrial Exhibition in Dublin. Once again, however, Bell's machine was judged superior.[51]

Despite the opinions of early observers, it was the American prototypes which were successfully developed by manufacturers. The English firm of Burgess and Key improved McCormick's reaper in 1854 by the addition of an Archimedean screw which delivered the cut grain to one side of the machine.[52] In 1859, these machines were being advertised in an Irish agricultural journal, with sales figures (presumably for both Britain and Ireland) and a warning that orders for the machines should be made early:

Numbers of Burgess and Key reapers sold
1856   50
1857   250
1858   700
Owing to the excessive demand for last season, many orders were left unexecuted. Burgess and Key, therefore, earnestly request that the orders for the ensuing season be sent as soon as possible.[53]

The advertisement also quotes a testimonial from several gentlemen who were present at a trial of one of the machines at Clandeboye, county Down, on 26th August, 1858. These gentlemen affirmed the effectiveness of the Burgess and Key machines on both 'a light crop of oats growing on a clean surface, and on a heavy crop with clover and grass growing thickly in the bottom'.[54]

Combined reaping and mowing machines were being manufactured by the late 1850s. These were powered by the large, heavy running wheels on their sides. The wheels were connected by gearing to a rod known as a 'pitman', which worked backwards and forwards. This movement was transmitted to the cutting knife of the reaper.[55] By the 1860s, agents and some manufacturers of reaping machines were dispersed throughout many parts of Ireland. At the Spring Show of the Royal Dublin Society, in April 1863, W. O'Neill of the Agricultural Implement Depot in Athy exhibited 'patent prize combined reaping and mowing machines'. Several Dublin exhibitors also had reaping machines on their stands. In August of the same year even more stands belonging to Irish firms selling reapers were exhibited at Kilkenny, including that of Thomas McKenzie of Munster Agricultural House, Cork. It is impossible to say from reports, however, which, if any, of the exhibitors were manufacturers rather than agents.[56] The largest Irish manufacturer of reaping machines was Pierce of Wexford, who probably began producing them in the early 1860s.[57] By the late nineteenth century, the firm was producing a variety of machines, most of which could be used for either mowing hay, or reaping grain.

In Britain, it has been estimated that by 1871 there were about 40,000 reaping machines, cutting as much as 25 per cent of the total grain crop. With a change of horses, a reaping machine could cut eight to ten acres in a day. Eight to twelve men were required for savings in cutting time to be taken advantage of, however. These men either tended the machine, or tied sheaves and stooked them. Overall, labour requirements were about 50 per cent lower than those required when scythes were used. However, although the Irish Industrial Exhibition catalogue estimated that as much as fifteen acres of grain could be cut by a reaping machine in a day, the editor was cautious about the immediate value of the machines for Irish farming. The probable cost of reaping fifteen acres by hand was compared to that for reaping a similar area by machine. It was concluded that using the machines could lead to a saving of 5s. 10d. per acre. This saving was judged important, but only 'where the extent of land under grain crops is considerable'.[58] The same saving would not be so significant for smaller Irish farmers, and when the price of the Burgess and Key machines quoted in 1859 (£42 10s.) is considered, the caution of Irish agriculturalists becomes understandable.

Apart from the greater capital input required to purchase a reaper, the machines presented Irish farmers, on a larger scale, with the same qualified advantages as the scythe. Harvesting could be achieved much more quickly, a major advantage in the changeable Irish climate, but more workers were required for this shorter period. By the end of the nineteenth century, reaping machines were common in the major tillage areas in Ireland (Fig. 86), but on the smallest farms, sickles and hooks remained efficient harvesting tools.

Fig. 86. A reaping machine in use near Toome, Co. Antrim, c.1920 (WAG 1158).

In 1887, the American firm, McCormick's, put a reaper-binder on to the market. This machine not only cut grain, but tied it into sheaves with string:

> A sheaf-binding harvester has four separate operations to perform — viz., cutting, elevating, binding and delivering the crop. In the binder the cutting apparatus differs only in details from the ordinary one-wheeled self-delivery reaper. The grain as cut falls across an endless web, which conveys it over the top of the driving wheel to the knotter, where the straw falls into two arms called compressor-jaws, which keep it on the knotter-table until a sheaf of any specified size has accumulated. Whenever a sheaf of the desired size has been delivered to the compressors, these relieve the tripper, which sets in motion the needle (carrying the binding twine) and the knotting apparatus. The needle is circular, and in its course it passes the band (twine) round the sheaf, when the band is caught by the knotter, and almost instantaneously a firm and secure knot is tied, while the needle is drawn back ready to operate on a new sheaf. As soon as the knot is tied and the string cut, the sheaf is ejected from the machine in a horizontal position, dropping on the ground on its side, quite clear of the machine.[59]

The machines were very soon for sale in Ireland, some being advertised as 'specially designed for Irish crops; will handle the tallest and heaviest grain'.[60] In 1908, the results of a test on a farm in Essex were published, which showed the savings which could be expected from using a reaper-binder:[61]

Reaper-binder — costs				Ordinary reaping machine — costs			
Six horses	£1	10	0	Four horses	£1	0	0
Two men	£0	14	0	Tying by hand, 6s. per acre	£3	12	0
Two boys	£0	7	0	Two men	£0	14	0
String, 2s.							
per acre	£1	0	0	One boy	£0	3	6
Oil	£0	2	0	Oil	£0	2	0
Total for 10							
acres	£3	13	0	Total for 12 acres	£5	11	6
Cost per acre, 7s. 3d.				Cost per acre, 9s. 3d.			

However, despite the effectiveness of reaper-binder design, and the obvious saving of labour, the spread of the new machines seems to have been slow before the First World War, 1914-1918. It was only than that the government's compulsory tillage policies made some large farmers invest in these expensive machines (Fig. 63). Once purchased, however, some machines had a long working life. Even after tractors replaced horses, reapers and reaper-binders were still used, either being simply pulled behind the tractor, or modified to be powered directly from the engine.

Sheaves, Stooks and Stacks

Before the widespread use of binding machines, sheaves were tied by hand, with 'corn-bands'. These bands could be made in many different ways, but a simple method, used widely in Ireland, was described and illustrated in *Stephens' book of the farm* (Fig. 87): 'The corn-band . . . is made by taking a handful of corn, dividing it into two parts, laying the corn-ends of the straw across each other and twisting them round . . . the twist acting as a knot'.[62] The sheaf was set in the middle of the band, the ends of which were then pulled around it, and twisted together. These twisted ends were then pushed under the band, and so also acted as a knot. If sheaves were to be left in the field for any length of time, it was important to have the grain heads in the bands pointing downwards, to allow rain water to drain off. If water collected in the heads, after a short time the seeds could germinate and sprout. Workers responsible for binding were also warned not to place the bands too near the stubble

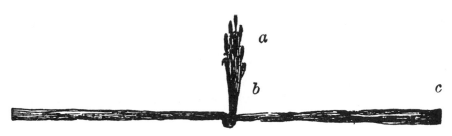

Fig. 87. A corn-band for tying sheaves (Stephens (1908), vol. 2, p.184).

Fig. 88. A 'gaitin' of oats (Stephens (1908), vol. 2, p.187).

ends of the sheaves as this made them stand less well in the stooks in which they were dried.[63]

The construction of stooks varied with weather conditions, and also the type of grain being harvested. Sheaves of grain, particularly oats, which had been bound while damp, were sometimes left standing singly

Fig. 89. A stook of oats or barley (Stephens (1908), vol. 2, p.187).

in the field in the same way as retted flax (Fig. 88). This was known in some areas as 'gaitin'. The bottom of the sheaf was spread out 'like a tent' to make it stand better.[64] The stooks most commonly described by nineteenth-century agriculturalists, however, were made up of two rows of sheaves, whose tops leaned on one another (Fig. 89):

> The binder, having bound up the sheaves, places them . . . resting upon their bases, and upon one another. Five pairs of these sheaves, when the crop is oats or barley, and six pairs when it is wheat, may be conveniently placed together in rows . . . The whole is covered by two sheaves, the butt ends of which lie towards each other, with the other ends divided a little, and pulled down so as to defend the upright sheaves.[65]

The two sheaves laid along the tops of the others were known as 'hooding' or 'head' sheaves. These protected the sheaves they covered from both rain and birds. Farmers were sometimes advised to remove the 'head' sheaves on fine days, and to turn the other sheaves around, so that the sides which had pointed in towards the centre of the stook would face the air. It was also recommended that where grain had been grown

Fig. 90. A stack of oats being built near Belmont, Co. Offaly, in 1969 (photograph courtesy of the National Museum of Ireland).

on ridges, the stooks should be built across the furrows. This would assist the circulation of air along the central line of the stooks.[66] Several agriculturalists advised that stooks should be aligned so that their ends faced north-south, as this would promote even drying.[67]

One curious omission from standard nineteenth-century Irish farming texts is the method of stooking described in the 1908 edition of *Stephen's book of the farm* as 'pirling':

> A plan of stooking sometimes pursued in certain exposed districts of the west and south-west is to set up two pairs of sheaves, the one pair at right angles to the other . . . When set up, the tops of the four sheaves are tied together about 9 inches under the apex, by a few straws pulled out of the top.[68]

There is plenty of photographic evidence that oats in particular were stocked in this way in Ireland (Fig. 84), and the technique is still used throughout the country when stooks are set up, yet the only evidence found so far that it may have been used in the nineteenth century is a comment in Stephens that 'pirling' was probably more common 'fifty years ago'. Farmers who still make these stooks point out that a major

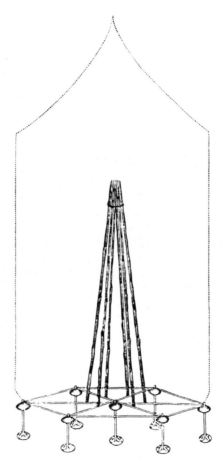

Fig. 91. A metal corn-stand, showing the outline of a stack built around it (*The farmer's gazette* (1844), vol. 3, p.216).

advantage is that they can be easily moved, both during and after drying. Since the sheaves are bound together, the whole stook can be lifted at once. This means that it can be easily shifted on to drier stubble. Lifting also allows some of the grass bound up with the sheaves to fall out.

Grain was left in stooks until the sheaves were judged to be dry: 'If perfectly dry, the sheaves will feel light and loose, and upon putting the hand into the heart no rawness will be perceptible'.[69] It was claimed by several agriculturalists that sheaves of grain cut with a scythe or a reaping machine were looser than those cut with hook or sickle, and so dried faster.[70] The length of time taken for drying, however, depended most on the weather. Some Irish farmers, especially in the midlands and southern counties, built sheaves into 'ricks', which were intermediate in

Fig. 92. Corn stacks on stone stands, near Athenry, Co. Galway, in 1970 (photograph courtesy of the National Museum of Ireland).

size between stooks and stacks, to allow complete drying.[71] However, it was a central aim to move the sheaves into grain stacks as soon as possible.

Most stacks in Ireland were built on a circular base. The sheaves were placed in layers, the stubble ends pointing outwards, and the grain heads meeting at the centre. Part of the skill in making a stack was to ensure that the sheaves also sloped slightly upwards towards the centre (Fig. 90). This meant that rain water seeping through the stack tended to drain outwards rather than stagnate around the grain heads, possibly leading to germination. Stacks were given conical tops, and these were often roughly thatched to protect the grain from rain, and tied down by ropes (often *súgán* ropes of twisted hay), to make the structure stable in windy weather.

Larger farms had a 'stack-yard' or 'haggard' in which the stacks were built, but on smaller farms an area in the corner of a yard or field was set aside for the purpose. Some of these smaller farmers made bases for their stacks, simply by laying furze or thorn-branches on the ground, but 'corn-stands' were also common. These held the stack off the ground and also discouraged a large-scale invasion by rats and mice. Stands varied widely in construction. By the mid-nineteenth century, iron stands were being produced by some local foundries (Fig. 91), while on some large farms elaborate stone bases were made. On the other hand, some smaller farmers constructed very crude stone bases from large flat pieces of local stone. Stands of dressed stone, shaped to resemble mushrooms, were also common (Fig. 92). The length of time the grain was stored in stacks depended on when threshing could be carried out, and which threshing technique was used, but once the stacks had been built, the worst worries and hardest labours of the grain harvest were over.

NOTES

1. Cullen, L.M., *An economic history of Ireland since 1600* (London, 1972), p.22.

2. *Ibid.*, p.46.

3. *Ibid.*, p.109.

4. Lambert, Joseph, *Observations on the rural affairs of Ireland* (Dublin, 1829), p.x.

5. Cullen, L.M., *op. cit.*, p.118.

6. Department of Agriculture and Technical Instruction for Ireland, 'Statistical survey of Irish agriculture', *Ireland industrial and agricultural* (Dublin, 1902), p.308.

7. Sproule, John, *A treatise on agriculture* (Dublin, 1839), p.285; Solar, Peter, 'Agricultural productivity and economic development in Ireland and Scotland in the early nineteenth century', in *Ireland and Scotland, 1600-1850*, eds, T.M. Devine and D. Dickson (Edinburgh, 1983), p.76.

8. Burroughs, E., *The Irish farmer's calender* (Dublin, 1835), p.216.

9. Baldwin, Thomas, *Introduction to practical farming* (Dublin, 1877), p.216.

10. *Belfast Newsletter* (Belfast, 3rd August, 1750).

11. Young, Arthur, *A tour in Ireland*, vol. 1 (Dublin, 1780), p.161.

12. Doyle, Martin, *A cyclopaedia of practical husbandry*, rev. ed. W. Rham (London, 1844), p.567.

13. Tighe, W. *Statistical observations relative to the county of Kilkenny* (Dublin, 1802), p.177.

14. Doyle, Martin, *op. cit.*, p.155.

15. Sproule, John, *op. cit.*, p.280.

16. *Ibid.*, p.279. .

17. Doyle, Martin, *op. cit.*, p.44.

18. Young, Arthur, *op. cit.*, vol. 1, pp.125, 141, 148.

19. This may have been the case in ancient Ireland. Joyce, P.W., *A social history of ancient Ireland*, vol. 2 (Dublin, 1920), p.273. For a summary of

international evidence, see Bell, J., 'Sickles, hooks and scythes in Ireland', *Folklife*, vol. 19 (Leeds, 1981), p.26.

20. Mitchell, Frank, *The Irish landscape* (London, 1976, p.152; Duigan, M., 'Irish agriculture in early historic times', *Journal of the Royal Society of Antiquaries in Ireland*, vol. 74 (Dublin, 1944), p.140.

21. Joyce, P.W., *op. cit.*, vol. 2, p.174.

22. 'Pairlement chloinne Tomáis', trans. O.J. Bergin, *Gadelica* (Dublin, 1912), p.43.

23. Howatson, W., 'Grain harvesting and harvesters', in *Farm servants and labourers in lowland Scotland*, ed. T.M. Devine (Edinburgh, 1984), p.129.

24. Loudon, J.C., *An encyclopaedia of agriculture* (London, 1831), p.373.

25. Doyle, Martin, *op. cit.*, p.493.

26. Davis, Francis, 'Low and clean', in *Earlier and later leaves* (Belfast, 1878), p.270.

27. Fenton, A., *Scottish country Life* (Edinburgh, 1976), p.54.

28. Roberts, Michael, 'Sickles and scythes: women's work and men's work at harvest time', *History workshop*, vol. 7 (Southampton, 1979), p.18.

29. O'Rahilly, T.F., 'Etymological Notes, 2', *Scottish Gaelic studies*, vol. 2 (Edinburgh, 1927), p.26.

30. Bell, J., *op. cit.*, p.29.

31. Young, Arthur, *op. cit.*, vol. 1, p.72.

32. Tighe, W., *op. cit.*, p.214.

33. Murphy, M.J., 'The mower from Moygannon', *Ulster folklife*, vol. 12 (Holywood, 1966), p.108.

34. Perkins, J.A., 'Harvest technology and labour supply in Lincolnshire and the East Riding of Yorkshire 1750-1800', *Tools and tillage*, part 2, vol. 3 (Copenhagen, 1977), p.128.

35. Ó Súileabháin, Amhlaiobh, *Cinnlae Amhlaiobh Úi Shúileabháin*, vol. 3, trans. M. McGrath (London: Irish Texts Society, 1930 (1936)), p.66.

36. Irish National Schools, *Agricultural classbook* (Dublin: Commissioners of National Education, 1868), p.155.

37. Dineen, Patrick S., *Foclóir Gaedhilge agus Béarla* (Dublin: Irish Texts Society, 1927), p.1095.

38. Bell, J., *op. cit.*, p.31.

39. Collins, E.J.T., *Sickle to combine* (Reading: Museum of English Rural Life, 1969), p.9.

40. Ó Súileabháin, Amhlaiobh, *op. cit.*, p.219.

41. Tighe, W., *op. cit.*, p.184.

42. Burroughs, E., *op. cit.*, p.265.

43. Roberts, Michael, *op. cit.*, p.16.

44. *Irish farmer's gazette* (Dublin, 29th August, 1863), p.297.

45. Perkins, J.A., *op. cit.*, p.130.

46. Doyle, Martin, *op. cit.*, p.155.

47. Fussell, G.E., *The farmer's tools* (London, 1952), p.115.

48. *Ibid.*, p.116.

49. *Belfast commercial chronicle* (Belfast, 1st October, 1806), p.2.

50. Fussell, G.E., *op. cit.*, p.129.

51. Sproule, John (ed.) *The Irish Industrial Exhibition of 1853: a detailed catalogue of its contents* (Dublin, 1854), p.216.

52. Fussell, *op. cit.*, p.132.

53. *The weekly agricultural review* (Dublin, 15th April, 1859), Supplement, p.viii.

54. *Ibid.*, p.viii.

55. Baldwin, Thomas, *op. cit.*, pp.316-317.

56. *Irish farmer's gazette* (Dublin, 1863), pp.132-133 and 301.

57. In the absence of contemporary documentary evidence, this claim is made on the basis of an article in the *Impartial reporter* (Enniskillen, 3rd August, 1967), p.2, that a Pierce No. 1 reaping machine in Ballybay, county Monaghan, had originally been bought in 1864.

58. Sproule, John, *op. cit.*, (1854), p.215.

59. *Stephens' book of the farm*, vol. 2, 5th ed., p.177.

60. Advertisement for McCormick harvesting machines, *The Dungannon news* (Dungannon, 1st May, 1902), p.2.

61. Stephens, *op. cit.*, p.181.

62. *Ibid.*, p.184.

63. *Purdon's practical farmer* (Dublin, 1863), p.283.

64. *Ibid.*, p.283.

65. Sproule, John, *op. cit.*, (1854), p.250.

66. Murphy, Edmund, *The Irish farmer's and gardener's register*, vol. 1, (Dublin, 1863), p.270.

67. Purdon, *op. cit.*, p.283.

68. Stephens, *op. cit.*, p.186.

69. Purdon, *op. cit.*, p.293.

70. Baldwin, Thomas, *The handy book of small farm management* (Dublin, 1870), p.84; Purdon, *op. cit.*, p.285.

71. Purdon, *op. cit.*, p.284.

CHAPTER 10

Threshing and Winnowing Grain

The most valuable components of grain crops, the seed and the straw, are usually separated before either is used. In Ireland, this has been achieved by a number of techniques, although by the eighteenth century some of these were regarded as exotic and primitive. One such technique which was used to obtain small quantities of seed involved burning the straw and husks, leaving the grain scorched, but otherwise undamaged. Burning might mean setting a whole sheaf alight, and then sifting the parched seed from the ashes. Alternatively, only the ears might be set alight and then knocked off the sheaf using a stick. A third method involved cutting off the ears from a sheaf and then burning them in a heap.[1] Burning was forbidden by an act passed in the Dublin parliament in 1634, but memories of bread made from burned grain (Irish: *loiscreán*) have persisted until very recently.[2] Arthur Young heard of the practice at Castlebar and Westport in county Mayo, but seems to have recorded it as a curiosity rather than as a common occurrence.[3]

One very simple way to remove seed from a sheaf was by lashing (Fig. 93). This is still sometimes practised on small farms if only a little grain seed is required, for example to feed chickens. Caiomhín Ó Danachair has summarised the technique as follows:

> The worker grasps the sheaf in both hands at the lower end and beats the ear end smartly against a suitably placed hard object, which may be a large stone, an upturned creel or tub, a stool or a chair, a barrel, a short pole supported horizontally, a ladder supported horizontally or leaning normally against a wall. The striking dislodges the grain, which falls to the ground; often a cloth or sheet is spread to catch it.[4]

Thatchers liked to work with straw which had been lashed, as the technique left it relatively undamaged. Wheat and rye, especially, were lashed as a preparation for thatching. In 1812, Wakefield found that throughout large areas of Ulster all of the wheat grown was lashed, although in this case the sheaves were also threshed (or 'thrashed') with a flail.[5]

The simplest form of flail was a single stick which was used to beat grain-seed out of sheaves. This technique was known in Ireland as

Fig. 93. Lashing oats on the Aran islands, Co. Galway, c.1930 (photograph courtesy of the National Museum of Ireland).

scutching, or in Irish, *scoth-bualadh*.[6] For many centuries, however, the most common type of flail (Irish: *súiste*) used in Ireland has been made from two sticks, of between three and four feet in length, tied together. One stick is held, and the other is used to beat the seed from the ears of grain. The different parts of flails are known by many different names throughout Ireland. Some of these are listed below:

> *Handstaff:* collop, *colapán, lamhchrann*
> *Striker:* buailteán, bóilcín, souple
> *Tying: iall,* thong, *gad,* hanging, tug, hooden, mid-kipple.[7]

The striker was usually made from a fairly hard wood, such as holly, blackthorn, whitethorn or ash, while the handstaff could be made from hazel, pine or oak. Ó Danachair has distinguished several types of Irish flails by differences in the way these sticks were joined (Fig. 94).[8] The tyings — which could be made from several sorts of animal skin (eel-skin being especially valued), scutched flax fibre, tough flexible withies of willows or (less desirably), rope — show some regional differences.

Flailing (Fig. 95) could be carried out by a man working alone, or up to four men working together.[9] If two or more men were threshing at the same time, it was important that each worked at a constant, complementary rhythm. The hard knocks received by flailers when this was not done are still remembered with amusement around the countryside. There was a basic difference in how flails were used, between those

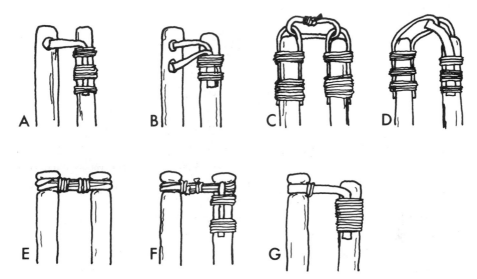

Fig. 94. Common Irish flail types: A. hole flail (Ulster) B. double hole flail (Ulster) C. cap-flail (Leinster, south Ulster, and east Munster) D. sub-type of cap-flail (Leinster and east Munster) E. double-loop flail (Munster, Connaught, and west Ulster) F. sub-type of double-loop flail (Munster) G. eye-flail (after Ó Danachair, 1971).

Fig. 95. Threshing with flails near Toome, Co. Antrim, c.1920 (WAG 296).

whose tyings allowed the striker to swing freely around the handstaff (Fig. 94e,f,g), and those where they could not do so (Fig. 94a,b,c,d,). When using the latter types, flailers let the handstaffs turn in their hands as they swung the implements. A man working alone could thresh between two and four sheaves at once. The sheaves were laid at the flailer's feet, and the ears, which were set to point in towards one another, were beaten with the flail. After a few strokes the sheaves were turned over, often using the end of the handstaff, and the other sides of the sheaves were then threshed.

There was some discussion amongst agriculturalists as to the best surface on which sheaves should be laid for flailing. In the early nineteenth century threshing often seems to have been carried out in the middle of the road.[10] Threshing in the open air continued on many small farms until recently, although it has been claimed that in Leinster it was almost always done in a barn.[11]

Agriculturalists, and farmers who can still use flails, agree that a good threshing surface should 'bend to meet the flail'.[12] Arthur Young disapprovingly recorded threshing floors made of clay, or at Furness, county Kildare, a mixture of lime, sand, and coal ashes.[13] The farmers who used these floors claimed that they did not damage the grain, but Young disagreed. He believed that the state bounty on grain should only be given for crops threshed on wooden floors:

> The samples of Irish wheat are exceedingly damaged by clay floors; an English millar knows the moment he takes a sample in his hand if it came off a clay floor, and it is a deduction in the value. The floors should be of deal plank two inches thick, and laid on foists two or three feet from the ground, for a free current of air to preserve them from rotting.[14]

On some large Irish farms, barns were constructed to make flailing as efficient as possible, rafters for example being specially raised to allow the striker to swing freely. On most farms, however, whether threshing was carried out in the barn, in the farm kitchen, or in the open air, the main preparation was to place a threshing board beneath the sheaves. This might be a door unhinged from an outhouse, or a smooth board which had been specially made for the purpose. Farmers liked to work on boards about six feet square.[15] A large cloth was usually spread on top of this, to collect the grain beaten from the sheaves.

In 1812, Wakefield claimed that on small farms in many parts of Ulster, grain was threshed almost as soon as it had been harvested, the main reason given being that on such farms there was no room for a haggard.[16] Later in the nineteenth century, however, flailing was a task given to labourers, or male members of the farming family, during bad weather or after dark in winter. The advantages of the flail were listed by Martin Doyle as 'its simplicity, the power of giving employment to the labourers in the barn during wet days, and the convenience of having fresh straw for fodder every day'.[17] Some attempts were made to calculate the amount a man could thresh in a day, using a flail. Doyle estimated that a good worker could thresh twenty stones of wheat, or thirty-two stones of oats or barley per day. Threshing machines, however, according to his calculations, could process more than twenty-five times this amount.[18]

Early Threshing Machines

Most of the early developments in the design of machines for threshing grain occurred in Scotland. Early inventions included an attempt to attach a series of flails to a revolving beam, but breakages were so common that this idea was abandoned. In 1786, however, Andrew Meikle of East Lothian produced a design which became basic to many later machines (Fig. 96). Here the grain was separated from the straw by a revolving cylinder, along with four bars of wood faced with metal had been fixed at regular intervals. Water power was used to turn the cylinder, or threshing drum as it became known, inside a metal casing set around the drum, about one inch away from its surface. Sheaves were fed through these, and the grain was beaten out of the sheaves by the bars on the cylinder. In later designs, revolving rakes were added, to rake off the straw. Meikle received an English patent for his machine in 1788, and from this time onwards threshers of his design began to be used on some large farms in Scotland and northern England.[19]

By 1800, threshing machines had been installed on some large Irish farms. A Mr. Christy of Kirkassock, near Magheralin, county Down, claimed to be the first proprietor in Ireland to have installed a thresher. In 1796, he had inspected machines in Scotland and had drawings and a model made of one of these. A local workman used these plans to construct a similar machine on Mr Christy's farm.[20] Christy's machine, and at least some other early installations, were driven by horses. Mr. Ward of Bangor, county Down, for example, had a 'horse engine', or gearing system around which the horses were driven in order to turn the thresher, set below the machine. This complex could be operated by two or three horses, and five workers. A boy drove the horses, while a man threw up the sheaves to be threshed to another worker, who unbound

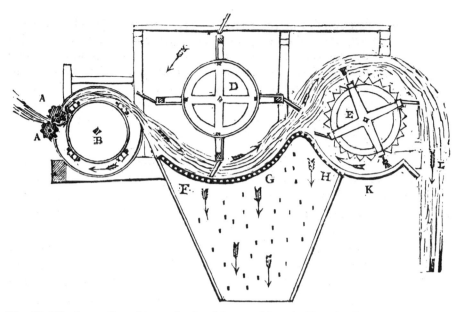

Fig. 96. The internal workings of a threshing machine: A. Fluted rollers, through which the sheaves of grain are passed into the machine. B. Cylinder with beaters attached to knock the seed from the grain. D,E. Drums with rakes attached, to shake the straw, and pull it forward through the machine. F,G,H,K. The sparred bottom of the machine, through which the grain seed and chaff fall. L. The opening through which the straw leaves the machine (Sproule, J. (1839), p.77).

them. A fourth man fed the sheaves into the machine, and the fifth man removed the threshed straw: 'With this force three times the quantity of grain is got out, besides its being half cleaned; but the greatest advantage is the facility with which a great quantity of grain is got ready for market in a comparatively short time'.[21] Another advantage cited was that, as with flails, threshing machines could provide inside work for labourers during bad weather.

Surveyors working for the Royal Dublin society reported that, by about 1800, threshing machines had been installed in several counties, including Cavan, Louth and Dublin. In the last county it was claimed that the machines were coming into use 'very fast'.[22] Machines continued to be installed on large farms during the first decade of the nineteenth century. By 1811, it was estimated that the cost of machines bought in Scotland and erected in Ulster by Scottish workmen had dropped to 50% of the 1800 price.[23] During the same period, locally made machines were being erected in county Cork. In 1813, for example, Mr. James Stopford of Brookville farm gave the following description of a machine installed on his farm:

The thrashing machine . . . is, I believe a very good one, though it does not (at least as yet) perform as many combined operations as others of which I have read the description. It is worked by an overshot, or what is more properly called a balance, waterwheel . . . The person who prepared and put it up is Francis Barry of Brooklodge, near Riverstown who (although no more than 28 years of age, as he informs me) has already erected on his own account since the year 1805, in the contiguous and even some more remote counties, about forty threshing machines.[24]

Despite this early activity, however, threshing machines seem to have been confined, during the first thirty years of the nineteenth century, almost entirely to farms where the farmers was wealthy enough to take on the expense of building a water-wheel, or a covered horse-engine, the two alternative sources of motive power for the machines.[25] The Ordnance Survey Memoirs recorded few threshing machines in the north of Ireland, even by the mid-1830s.[26] The situation reported from Lavagh parish, county Cavan, seems to have been typical: 'The flail and the centre of the road answer every convenience of the threshing machine and barn'.[27] More widespread adoption of mechanical threshing only came when smaller machines became available.

Types of Threshing Machine

Machines powered by water were usually large, and often formed part of a complex which might also include a winnowing machine, and even corn grinding stones. Although, as we have seen, horses were used to power even the earliest machines, agriculturalists advised against this: 'When other means can be used, horses should not be employed for the purpose, the labour being severe, and the time thus occupied being taken from that for performing the other labours of the farm'.[28] Horse gears seem to have put an uneven strain on the animals, and the circular motion of turning them could cause giddiness.[29] Martin Doyle believed that a contrivance invented by the Scot, Walter Samuel of Niddry, which meant that horses pulled with a continuous and equal draught, would lessen the problem,[30] but as late as 1863 *Purdon's practical farmer* described the use of horses for threshing as 'most objectionable'.[31] Between two and six horses were used for driving the earliest machines, but by the 1840s much lighter horse engines had been developed. In 1842, William Neill of the Hibernian foundry in Belfast was advertising himself as the inventor of a one horse-power threshing machine.[32] Whatever the truth of this claim, the smaller, lighter horse-engines

SHERIDAN'S THREE-HORSE POWER THRASHING MILL.

Fig. 97. A threshing machine, and gearing system turned by three horses, manufactured by Sheridan of Dublin (*The farmer's gazette* (1848), vol. 7, p.398).

available from this time were a significant inducement to farmers to install threshers. The new gearing systems made use of cast metal. The horses were still harnessed to the outer end of a long arm (or arms), which turned around a central mechanism, but in the new forms the driving shaft which connected the horse-engine to the threshing machine came out from beneath the central gearing (Fig. 97). In the most common arrangement, the driving shaft was often encased in a stone-lined channel sunk in the ground.[33] The barn thresher which was operated by the new gearing systems also became increasingly standardised (Fig. 98).

In 1837, the firm of Pierce began to make machines in Wexford, and soon became the largest manufacturer in Ireland.[34] By the 1860s, many smaller foundries were also making machines. In 1863, there were four manufacturers based in Belfast alone, but the northern foundry producing most machines was Kennedy's of Coleraine which began making threshers in 1860. R.A. Gailey has traced sales of threshing machines made by this company, and has found that between 1860 and 1940 about 2,000 machines were produced and installed, mostly in the north-east, but also in western counties such as Fermanagh and Donegal.[35] Using documents preserved by the company, Gailey has also been able to show the importance of examples set by neighbours for the spread of threshing machines. Letters to the firm make this explicit. In one example, a county Tyrone farmer, William Moorehead of Ballymullarty, Newtonstewart, wrote asking that Kennedy's should send a machine

as soon as you can as I would like to have her up before harvest. I expect she will be worthy of the praise you give your mills as I have some

Fig. 98. A horse-powered threshing machine on an Ulster farm, c.1920 (WAG 1959).

neighbours anxious to see her who wants one. You can let me know when to look for her at Victoria Station [Victoria Bridge] and when you will send the man to put her up.

This particular letter also illustrates the point that the development of a rail network in the 1840s and 1850s was probably significant in facilitating the spread of these heavy machines.[36]

Barn threshing machines were made in several sizes. The dimension usually quoted in advertisements was the width of the threshing drum. Thus Kennedy's at one time manufactured eleven different-sized drums, ranging from 18 inches to 54 inches in diameter.[37] There was also some discussion in farming journals of the mid-nineteenth century of more basic structural differences between machines, however. While many machines continued to be fitted with bars or beaters of the kind devised by Andrew Meikle, others had drums with rows of short iron pegs to beat out the grain. In 1849, both 'beater' and 'peg' drums were being manufactured in Cork.[38] In England it was claimed that 'beater' drums damaged the grain, although it was also argued that the damage was more the result of the speed at which machines were worked.[39] In Ireland, on the other hand, there was some discussion as to whether or not 'peg' drums broke up straw so much that it was useless for thatching.[40] Threshing machines with both types of drum continued to be made, but for larger machines the beater drum seems to have been more commonly adopted.[41]

All threshing drums turned at speeds which could make them dangerous to work with, and also mean that weak parts of badly constructed machines were liable to break. One correspondent of the *Irish farmer's gazette* warned that peg drums were particularly dangerous:

> A chance stone in a sheaf, striking against the pegs, will sometimes break some of them in the middle of the thrashing, and driving them with the force of a pistol ball, may do serious mischief should any person be in their way . . . to guard against injury to life or limb from such accidents, everyone about this machine is cautioned not to expose himself by standing opposite the delivering face of the drum.[42]

The problem of damage to machines could be lessened, according to John Sproule, by building up the speed of turning gradually. After top speed had been reached the corn should be fed in regularly, a rate of one sheaf at a time being sufficient for most machines. Grain should not be put in to such an extent that it slowed down the turning of the threshing drum. This was to be watched for particularly if oats with long, damp straw were being threshed.[43]

We have already seen that the smaller sizes of barn threshing machines produced in the mid-nineteenth century meant that threshers became a viable investment for some less well-off farmers. However, two other developments occurring around the same time meant that even very small farms were able to make use of the technology. These developments were the manufacture of portable threshers, and of very small machines, turned by hand (Fig. 99). Hand-operated threshing machines were known in Britain by the early nineteenth century. By 1813, they had made their appearance in Ireland:

> The Dublin Society has lately got a threshing machine constructed to be worked by four men; two of them are employed in turning the machine; and the other two are engaged in feeding and carrying away the corn and straw. It is capable of threshing 20 barrels of oats per day or 16 of wheat. It can easily be removed from one place to another, and can be worked in a room 12 feet square.[44]

This early introduction does not seem to have made a great impact however, as in 1849 *The Irish farmer's gazette* was enthusiastically reporting the exhibition of another hand-threshing machine in the Royal Dublin Society's agricultural museum as a new solution to the problem of the expense of buying a thresher. The rate of working of this later

Fig. 99. A hand-operated threshing machine in use in the Ulster Folk and Transport Museum (UFTM, L1531/1/3a).

machine — twelve barrels of oats, and eleven barrels of wheat per day — was less than that claimed for the earlier machine, but the makers, Barrets of Reading, Yorkshire, supplied many machines to Irish farmers during the second half of the nineteenth century.[45] Machines which could be worked by pedals as well as being turned by hand also became available, and early in the present century some of these were supplied to small farmers in those western counties known as the 'congested districts'. These machines cost only £7 each, and if five men were working them, could thresh between twenty-two and twenty-four stones per hour.[46]

The other solution widely adopted to overcome the problem of the still relatively high purchase cost of barn threshers was to hire a portable threshing machine. The development of lighter cast-iron gearing systems meant that it became possible to move the entire threshing system from farm to farm. A horse-operated portable thresher was patented in England in the 1840s, and shown at the Great Exhibition of 1851 in London.[47] In 1853, similar machines were shown at the Irish Industrial Exhibition in Dublin. John Sproule expressed some concern that this type of machine would be prone to breakages during movement. He accepted, however, that well-made machines, such as that exhibited by Ransomes and Sims of Ipswich, would be strong enough to take this

DUNDALK SHOW,

JULY 27th to 29th.

RICHARD GARRETT & SONS,
LEISTON WORKS, SAXMUNDHAM, SUFFOLK,

BEG respectfully to intimate to their numerous Friends, and Agriculturists generally, that they purpose exhibiting a varied assortment of their *Approved Machines* at the above Meeting, including several new and improved Implements, some of which will be shown in operation. Amongst the novelties exhibited by them will be found the newly-invented "Fixed and Portable Combined Finishing Threshing and Dressing Machines."

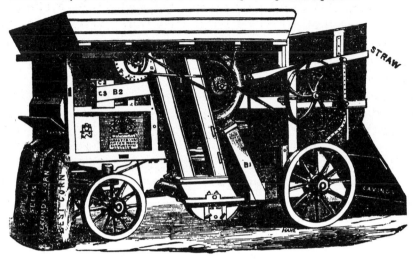

Fig. 100. A steam-powered threshing machine, manufactured by Garrett of Suffolk, and exhibited in Dublin in 1853, and Dundalk in 1859 (*The agricultural review* (1859), vol. 1, p.xcviii).

strain.[48] He also pointed out very clearly the potential value of portable machines for small farmers:

> There is no room for doubt as to the value under certain circumstances of the portable form of threshing machine. Through its instrumentality the use of machinery may be available on many farms, which must otherwise be satisfied with the flail. In some two or three days, the contents of a small stack-yard may be got over, when the machine may be taken elsewhere to be similarly employed. To the circumstances of a large proportion of Ireland portable threshing machines are peculiarly applicable, and no small service would be rendered by proprietors in keeping them for the use of their tenantry. They might even be profitably turned to account by some enterprising farmer in every locality hiring out a machine to be superintended by his own servant, and worked, either by his own horses, or that of the farmer requiring the work done, as might be mutually arranged between the parties.[49]

In the 1860s portable threshing machines were claimed to be 'partially' used in Ireland,[50] and horse-operated portable threshers continued in use well into the present century.[51] Portable machines became much more widely used, however, when steam power was applied to threshing. A Lincolnshire firm, Tuxfords of Buxton, were claimed to be the first to develop this combination, in 1842.[52] Another English maker, Garrett of Suffolk, exhibited a portable steam threshing complex in Dublin in 1853 (Fig. 100).[53] A machine made by this company was successfully tested on the farm of a Mr. Christy of Adare, county Limerick, in 1859,[54] and during the next eighty years the machines became increasingly popular. One effect of their increasing use was probably to lower sales of fixed equipment, especially after the First World War.[55] Some richer farmers followed the strategy outlined by Sproule, and hired out machines around the countryside. In one exceptionally large operation, early in the present century, a county Derry farm hired out at many as fourteen threshing machines at one time.[56] A common arrangement seems to have been that the steam thresher and its operator would stay on the farm until its entire grain crop had been threshed. As well as paying for the hire of the machine, the farmer would be expected to provide the operator with board and lodging. He might also be expected to refill the coal bunker of the steam engine before it left the farm.

A large supplementary labour force was required to take full advantage of the speed at which steam threshing was carried on. Long-established relations of mutual help between neighbours (Irish: *meitheal*) were utilised to provide this. An account from near Bailieboro, county Cavan, describes the division of labour of the workers:

1. Two men on top of the reek or stack who pitched sheaves on to the platform of the drum of the thresher.
2. A couple of men or boys would take these and cut off the tying bands before passing them to the 'feeder'.
3. The 'feeder's' job was to feed the sheaves into the machine.
4. One or two men would be kept busy filling bags with grain and carrying them to the barn.
5. Meanwhile, the straw coming out at the back of the drum of the machine was pitched by more workers to those building the stack.[57]

This use of co-operative working relationships between neighbours provides one of the clearest examples of technological change actually strengthening these links. A lot of writings dealing with 'modernisation' tend to assume that changing techniques mean less mutual help.

H

Hummelling

Barley presented particular problems during threshing. The heads of
this crop have long spines or 'awns', which could not be removed
efficiently by either flails or most unmodified threshing machines.
Removing the awns could require a separate operation, which became
known as hummelling, awning, or avelling.[58] Implements known as
hummellers were devised for this (Fig. 101). One common type had a
shaft with a T-shaped handle, and at the other end a square metal plate
made of parallel strips of iron. The barley was spread on the ground and
hit with the hummeller, until the awns were chopped off.

Threshing machines could be modified so that they hummelled barley
as part of the threshing process. Fluted covers were manufactured which
could be fitted over the threshing drum.[59] In 1839 *The Irish farmer's
gazette* suggested a simpler modification, however:

> We are not aware that our Irish manufacturers keep [hummelling covers]
> . . . ready made, but doubtless any of them could furnish you with one by
> ordering it. If you have a thrashing machine, by taking the cover off the
> drum, and putting on another lined with tin, perforated with small holes,
> three sixteenths of an inch wide, by passing the barley through the rollers,
> after being thrashed, the awns will be rubbed off in a very short space of
> time.[60]

In the same year, a peg-drum threshing machine made in Cork was
claimed to remove barley awns without any modification being
necessary.[61] This was said to be due to the velocity at which the machine
was driven. As we have seen, however, very high speeds of working were
declared unsafe by many agriculturalists.

The major advantage claimed for threshing machines was that they
were fast. One of the more unusual benefits claimed as a result of this
speed was that it 'saves pilfering, [the grain] . . . not being kept too long
under the hands of workmen'.[62] However, although the same writer
claimed that threshing machines were much more effective than flails in
beating grain out of sheaves, their overall efficiency was not un-
questioned. In 1813 *The Munster farmer's magazine* advised that unless
sheaves had been cut neatly with a sickle, so that they could be fed into a
machine very evenly, each sheaf should be threshed twice.[63] Also, as has
already been mentioned, sheaves with very long straw could slow the
machine down, and agriculturalists advised that long straw should be cut
before threshing. This had to be done carefully, however, since if the

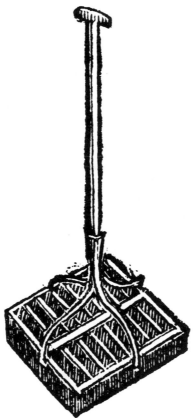

Fig. 101. A hummeller, for removing spines or 'awns' from the barley grain (Sproule, J. (1839), p.280).

straw was cut too short the sheaves might pass through the machine too quickly to be properly threshed.[64]

Worries were also expressed about the general efficiency of machines' construction. In 1854, John Sproule wrote

> Year after year we . . . [have found], that even the best manufacturers continued to turn out machines, the three-fourths of the power to work which was necessary for merely putting them in motion. Thus the four-horse power threshing machines actually required three horses to work them when empty, leaving to the fourth the whole labour of threshing the corn.[65]

The calculations of costs involved in threshing with either flails or threshing machines were complex. The earliest machines were large and expensive, whereas Irish farm labour was so cheap that even in the mid-nineteenth century it was not always clear whether it was worthwhile for an affluent farmer to purchase a threshing machine. In 1849, for example, a correspondent in the *Irish farmer's gazette* asked if it was more economical to purchase a threshing machine than to have oats threshed with flails, which he could have carried out for the cost of 3½d per barrel of seed produced. The *Gazette* replied that it would certainly be much cheaper to thresh with a machine. This was qualified, however, by some remarks on the proper use of labour. We have seen that both threshing machines and flails were seen as useful means by which a farmer could keep labourers working during bad weather, or after dark. This was usually presented in terms of the farmer getting maximum returns from his labourers, especially if they were hired servants living on the farm. The *Gazette*, however, echoed the call of English farm workers who opposed the introduction of threshing machines in that country: 'In order to give employment, we would prefer thrashing [with flails] at 3½d a barrel than using a machine . . . at the same time we are under the impression that 3½d a barrel is very poor hire'.[66] The declining number of farm labourers in the late nineteenth century, and the methods described above whereby even small farmers could avail themselves of the services of a threshing machine, meant that the latter became increasingly dominant. However, once again it is clear that in comparing 'common' and 'improved' techniques, we cannot simply oppose them to one another as efficient and inefficient. Many other factors must be considered before the rationality of decisions to retain old techniques, or adopt new ones, can be assessed.

Winnowing

After threshing, the husks or shells around grain seed were removed. For most of the eighteenth century, and indeed into the twentieth, this was done out of doors on a breezy day by simply pouring grain from a container (Fig. 102). The heavy grain fell straight down on to a sheet laid out for the purpose, while the light husks were blown away. This open-air winnowing could take place either on a road or in a field. Arthur Young noted instances of both,[67] as did Tighe in county Kilkenny in 1802.[68] In early nineteenth-century Ulster, winnowing was claimed to be women's work.[69] High windy places were particularly valued as winnowing places. These were sometimes known as 'Shilling' (shelling) hills:[70]

Fig. 102. Winnowing grain on the Aran islands, Co. Galway, by pouring it from a skin tray or *bodhran* in a breeze, so that the heavy seed falls to the ground, and the light chaff is blown away, c.1930 (photograph courtesy of the National Museum of Ireland).

> After [grain] . . . is threshed, they convey it up to the edges of the hills, where a number of persons are employed in exposing it to the wind with uplifted arms, to separate the corn from the chaff. The effect of this process, when seen at evening, the winnowers being at the edge of the horizon, is curious.[71]

The grain to be winnowed was shaken from riddles, sieves or circular trays. Riddles were also used to clean grain before winnowing. Throughout the nineteenth century, and well into the present one, local experts made riddles from wooden strips (usually of ash), woven and attached to a circular frame. The size of the meshes or holes in the riddles was varied according to what type of grain was to be cleaned. One maker, whose techniques were recorded in county Dublin during the 1940s, made riddles with meshes of the following dimensions:

> Barn riddle — meshes 25 x 32 mms.
> Oats riddle — meshes 6 x 9 mms.
> Barley riddle — meshes 10 x 6 mms.[72]

This craftsman also made winnowing trays, known as 'blind sieves', since the wooden strips were closely woven together so that there was no space between them. The Irish name for this type of tray, *dallán*, was also given to winnowing trays made from sheep or goat skin stretched around a circular wooden frame. Skin trays like these were found throughout Ireland. On some western islands where wood was scarce, the rims were made from rings of straw rope (Irish: *súgán*).[73] The names given to the trays varied. In the north a smaller version of the *dallán* was known as a wight or wecht, while in the south the term *bodhran* or borrane was used.[74] *Bodhran* is also the name given to a one-sided Irish drum which is very popular with Irish folk musicians.[75] It seems probable that before the modern musical development of the *bodhran*, it had the dual functions of tray and drum.

Barn Fans

James Meikle, the father of Andrew Meikle whose work on developing threshing machines was mentioned earlier, developed a machine for winnowing grain, based on designs he had seen in Holland. Meikle's winnowing machine or 'barn fan' set up a current of air by turning canvas sails. As grain was fed into the machine, it fell through this artificial breeze which separated the seed from the husks. The seed was also cleaned by passing through a series of sieves. In 1770, the Englishman James Sharp constructed a more compact version of the machine. In this type the draught was created by wooden fans, turned by a handle on the side of the machine's wooden casing[76] (Fig. 103). The simple construction of these barn fans made them cheaper than threshing machines, and it has been suggested that barn fans were adopted in Ireland a generation or more before threshers.[77] By the beginning of the nineteenth century, barn fans were claimed to be in common use around Dundalk, for example.[78] The principle behind the machines was easily grasped. This was amusingly illustrated by the Rev. Gerald Fitzgerald of the parish of Ardstraw, county Tyrone:

Fig. 103. A winnowing machine manufactured by Sheridan of Dublin (*The farmer's gazette* (1849), vol. 8, p.494).

> Soon after I came to this parish, in 1806, I got a winnowing machine, and having on a calm day, during its operation in the barn, asked a man who used before to winnow in the field, how he liked it? He exultingly replied, 'O sir, we need not now be looking abroad for the wind, we have her under lock and key!'[79]

Barn fans were easy to use. Because of the gearing system, the fans could be turned for a long time, by a woman or a boy, without difficulty.[80] By the mid-nineteenth century fans not only separated the grain from sticks and stones, but also delivered grain of different weights, and therefore different quality, from different spouts. Wire-meshed riddles of different sizes were fitted to the machine depending on which type of grain was being winnowed. Wheat riddles had five meshes to the inch, barley riddles four meshes to the inch, and oats riddles three meshes to the inch. A well-made machine would winnow or 'dress' oats in one go, but farmers were advised that wheat and barley grain should always be put through the machine twice.[81]

In large early nineteenth-century complexes, winnowing machines were commonly placed on the lower floor of a barn, with a threshing machine placed on the floor directly above. This meant that newly threshed grain was easily passed on for winnowing. By the 1830s, however, American manufacturers were producing machines which threshed and winnowed grain as parts of a single operation.[82] Steam-powered threshing machines such as Garrett's (Fig. 100) also bagged the grain. Hand-threshers and barn-threshing machines usually did not perform these extra functions, however, and on farms on which these were used, winnowing remained a separate activity.

Grain Storage

Some agriculturalists advised farmers not to thresh their entire grain crop at once. In 1849 *The Irish farmer's gazette* discussed the relative losses of grain kept unthreshed in stacks, and that stored in a heap after being threshed. The *Gazette* emphasised that the amount of grain lost from a stack depended very much on the dryness of the site on which it had been built, but estimated that the loss in general would be between 10 and 20 per cent. However, it was claimed that this was a small amount compared to that lost from threshed grain. In a stack, the *Gazette* argued, 'the grain will be kept sweet, if properly saved, and the only loss will arise from evaporation; but if thrashed, and kept over, the corn is very rarely kept sweet . . . so that it is a practice that should not be adopted'.[83] We have seen, however, that Irish farmers, whether depending on flails or large threshing machines, tended to thresh their grain crops as quickly as possible. Where grain was grown for cash, the crop was often sent to market almost immediately.[84] Grain-seed kept on the farm, however, was stored in a variety of ways. In 1802, the lack of proper storage on some county Kilkenny farms was severely criticised by Tighe: 'In unfavourable weather corn sometimes lies exposed in a hovel called a barn, to vermin and damp'.[85] Field-workers in county Cork have recorded an assertion that grain which was stored unwinnowed was particularly vulnerable to attacks by rats and mice: the husks of the grain made the mass more compact, allowing them to tunnel into the heap. Dry, winnowed grain formed a much more mobile mass, and could not be tunnelled into so easily.[86] Presumably, the experts of *The Irish farmer's gazette* who advised against early threshing had in mind heaps of unwinnowed grain kept in the way observed and condemned by Tighe.

This quality of storage must be contrasted with other evidence, however. Arthur Young, for example, gave high praise to a granary he examined on one large farm in county Antrim: 'Mr Lesly's granary is one of the best contrived I have seen in Ireland; it is raised over the threshing floor of his barn, and the floor of it is a hair-cloth for the air to pass through the heap, which is a good contrivance'.[87] Some small farmers also stored grain carefully. One method, recently recorded in counties Cork and Kerry, involved the construction of a cylindrical granary using very thick coils of straw rope (Irish: *súgán*). The base of the granary could vary from between 6 feet and 26 feet in diameter. Circular stands made from slabs of stone were sometimes used to hold the base off the ground, but in any case a protective layer of furze, twigs or brambles was laid down before being covered by a layer of straw. The sides were built up from coils of straw rope, grain being added as the work progressed. Recorded examples had sides between 5 and 6 feet high. The construction was completed with a domed roof of thatch which brought the overall height to between 7 and 12 feet.[88] Lucas has argued that this type of granary (known by several names, including *fóirín*, *síogog*, *sop*, *dárnaol* and *lochtán*) has been made in Ireland for at least several hundred years, and probably once had a much wider distribution throughout the country. Since the construction was filled as it was being built, this may have provided a further incentive for farmers to thresh their crops quickly.

Threshing, winnowing and storing grain crops were closely related operations. The method which a farmer used depended on how quickly he wanted grain, his ability to gain access to technological developments, and how he decided to utilise his labour supply. It is only by taking all these considerations into account that we can understand not only how single operations were performed, but also how these were related to other aspects of the processing of harvested grain.

NOTES

1. Ó Danachair, Caiomhín, 'The flail and other threshing methods', *Journal of the Cork Historical and Archaeological Society*, vol. 60:19 (Cork, 1955).

2. *Ibid.*, p.7.

3. Young, Arthur, *A tour in Ireland*, vol. 1 (Dublin, 1780), pp.350 and 364.

4. Ó Danachair, Caiomhín, *op. cit.*, p.8.

5. Wakefield, E., *An account of Ireland, statistical and political*, vol. 1 (Dublin, 1812), p.364.

6. Ó Danachair, Caiomhín, *op. cit.*, p.9.

J

7. *Ibid.*, p.12; Evans, E.E., *Irish folkways* (London, 1957 (1967)), p.213.

8. Ó Danachair, Caiomhín, 'The flail in Ireland', *Ethnologia Europaea*, vol. 4 (Arnhem, 1971).

9. Ó Danachair, Caiomhín, *op. cit.* (*1955*), p.9.

10. Wakefield, E., *op. cit.*, vol. 1, p.364; Binns, J., *Miseries and beauties of Ireland*, vol. 1 (London, 1837), pp.265-266; *Ordnance Survey Memoirs*, box 19, 7 (1835).

11. Ó Danachair, Caiomhín, *op. cit.* (1955), p.9.

12. Evans, E.E., *op. cit.*, p.215.

13. Young, Arthur, *op. cit.*, vol. 1, pp.218 and 231.

14. *Ibid.*, vol. 2, 'General observations', p.144.

15. Ó Danachair, Caiomhín, *op. cit.* (1955), p.9.

16. Wakefield, E., *op. cit.*, vol. 1, p.364.

17. Doyle, Martin, *A cyclopaedia of practical husbandry*, rev. ed W. Rham (London, 1844), p.553.

18. *Ibid.*

19. Partridge, Michael, *Farm tools through the ages* (Reading, 1973), p.162.

20. Gailey, R.A., 'Introduction and spread of the horse-powered threshing machine to Ulster's farms in the nineteenth century: some aspects', *Ulster folklife*, vol. 30 (Holywood, 1984), p.40.

21. Dubordieu, John, *Statistical survey of the county of Down* (Dublin, 1802), pp.53-54.

22. Archer, Joseph, *Statistical survey of the county of Dublin* (Dublin, 1801), p.44; Gailey, R.A., *op. cit.*, p.40; Beaufort, D.A., 'Materials for the Dublin Society agricultural survey of county Louth', ed. C.C. Ellison, *Journal of the county Louth Archaeological and Historical Society*, vol. 18:1 (Dundalk, 1973), p.124.

23. Gailey, R.A., *op. cit.*, p.40.

24. *The Munster farmer's magazine*, vol. 3 (Cork, 1814), p.34.

25. Gailey, R.A., *op. cit.*, p.39.

26. *Ibid.*, p.41.

27. *Ordnance Survey Memoirs*, box 19: 8, 7, 13 (1835).

28. Sproule, John, *A treatise on agriculture* (Dublin, 1939), p.78.

29. Fussell, G.E., *The farmer's tools* (London, 1952), p.162.

30. Doyle, Martin, *op.cit.*, p.553.

31. *Purdon's practical farmer* (Dublin, 1863), p.296.

32. *Northern Whig* (Belfast, 7th May, 1842), p.3.

33. Gailey, R.A., *op. cit.*, p.45.

34. Pierce Ltd., *The story of Pierce* (Wexford, 197?), p.1.

35. Gailey, R.A., *op. cit.*, p.47.

36. *Ibid.*, p.50.

37. Advertisement for Kennedy Ltd. (Coleraine: Kennedy, 19?).

38. *The farmer's gazette*, vol. 8 (Dublin, 1849), p.63.

39. Fussell, G.E., *op. cit.*, p.166.

40. *The farmer's gazette*, vol. 8 (Dublin, 1849), pp.63 and 81.

41. *Ibid.*, p.63.

42. *Ibid.*

43. Sproule, John, *op. cit.*, p.78.

44. *The Munster farmer's magazine,* vol. 2 (Cork, 1813), p.356.

45. *The farmer's gazette,* vol. 8 (Dublin, 1849), pp.463 and 474.

46. Congested Districts Board for Ireland, *Twelfth annual report* (Dublin: H.M.S.O., 1903), p.13.

47. Fussell, G.E., *op. cit.*, pp.164-165.

48. Sproule, John (ed.) *The Irish Industrial Exhibition in 1853: a detailed catalogue of its contents* (Dublin, 1854), p.21.

49. *Ibid.*, p.219.

50. Purdon, *op. cit.*, p.298.

51. O'Neill, Timothy P., *Life and tradition in rural Ireland* (London, 1977), plate 99.

52. Purdon, *op. cit.*, p.298.

53. Sproule, John (ed.), *op. cit.* (1854), p.224.

54. Purdon, *op. cit.*, p.299.

55. Gailey, R.A., *op. cit.*, p.80.

56. Interview with J.C. Drennan (Ulster Folk and Transport Museum tape C77.116).

57. O Dowd, Anne, *Meitheal: A study of co-operative labour in rural Ireland* (Dublin: Comhairle Bhéaloideas Éireann, 1981), p.102.

58. Purdon, *op. cit.*, p.300.

59. *Ibid.*

60. *The farmer's gazette,* vol. 8 (Dublin, 1849), p.474.

61. *Ibid.*, p.105.

62. Archer, Joseph, *Statistical survey of the county of Dublin* (Dublin, 1801), p.44.

63. *The Munster farmer's magazine,* vol. 2 (Cork, 1813), p.166.

64. *The farmer's gazette,* vol. 7 (Dublin, 1848), p.30.

65. Sproule, John, *op. cit.* (1854), p.218.

66. *The farmer's gazette,* vol. 9 (Dublin, 1849), p.342.

67. Young, Arthur, *op. cit.*, vol. 1, pp. 36 and 74.

68. Tighe, W., *op. cit.*, pp. 215-216.

69. Wakefield, E., *op. cit.*, vol. 1, p.364.

70. Evans, E.E., *op. cit.*, p.213.

71. Binns, J., *op. cit.*, vol. 1, pp.265-266.

72. Lucas, A.T., 'Making wooden sieves', *Journal of the Royal Society of Antiquaries of Ireland,* vol. 81:2 (Dublin, 1951), pp.147-148.

73. Evans, E.E., *op. cit.*, p.211.

74. Hall, Mr. and Mrs. S.C., *Ireland: Its scenery, character, etc.*, vol. 1 (London), p.83.

75. Ó Súilleabháin, Michael, 'The bodhran, parts 1 and 2', *Treoir,* vol. 6 (Dublin, 1974).

76. Partridge, Michael, *op. cit.*, p.165.

77. Gailey, R.A., *op. cit.*, p.39.

78. Beaufort, D.A., *op.cit.*, p.129.

79. Mason, William Shaw, *A statistical account, or parochial survey of Ireland*, vol. 1 (Dublin, 1814), p.129.

80. Sproule, John, *op. cit.* (1839), p.82.

81. Purdon, *op. cit.*, p.299.

82. Partridge, Michael, *op. cit.*, p.164.

83. *The farmer's gazette*, vol. 8 (Dublin, 1849), p342.

84. Binns, J., *op. cit.*, vol. 1, pp.265-266.

85. Tighe, W., *op. cit.*, p.216.

86. Lucas, A.T., 'An fhóir: a straw rope granary', *Gwerin*, vol. 1 (Oxford, 1957), p.3.

87. Young, Arthur, *op. cit.*, vol. 1, p.211.

88. Lucas, A.T., *op. cit.*, (1957); Lucas, A.T., 'An fhóir: a straw rope granary: further notes', *Gwerin*, vol. 2 (Oxford, 1958).

CHAPTER 11

The Theory and Practice of Improvement

Some recent historical writing has begun to question the established view that Irish agriculture during the last three hundred years was backward and inefficient. Levels of production of both crops and livestock have been used as evidence that, by the early nineteenth century, Irish systems of agriculture could stand comparison with those of England and Scotland, and certainly with those of north-west Europe.[1] The material presented in this book can be used to support this position. First, however, we should consider contemporary criticisms of Irish farming practices. We will then examine these in the light of how implements and techniques were actually used. In doing this the main aim will be to show the rationality of Irish farming methods.

During the later eighteenth and early nineteenth centuries, almost every general account of Irish cultivation techniques emphasised their defects. Common complaints were that rotations were minimal, fields weed-infested, and the land frequently cropped to exhaustion. Tighe's summary, published in 1802, is typical:

> that mode of cultivation is, indeed, too well known in Ireland, by which a soil naturally fertile, but exhausted by repeated crops of corn, is abandoned to noxious weeds for several succeeding years: is again broken up, slightly manured, exhausted, and again abandoned: where culture, instead of improving, deteriorates; where no effort is made for permanent utility. . . and where the different branches of rural economy, so far from assisting each other, remain unconnected and distinct, in a state of natural repulsion.[2]

Agricultural implements were also dismissed as inadequate. A typically sweeping condemnation was made by Loudon in 1831: 'The agricultural implements used in Ireland are all of the rudest construction. The plough, the spade, the flail, the car, all equally partake of imperfections and defects'.[3]

In their criticisms, agriculturalists tended to contrast the backwardness of Irish cultivation techniques with those currently advocated, and to some extent implemented, on English and Scottish farms. This 'improved' farming was believed to arise from the systematic application

229

of ideas derived from the science, or sometimes the art, of agriculture. There was some recognition that, as with any successful experimentation, the theory and practice of agriculture were best developed by interaction, but it was only rarely conceded that Irish farming was also advanced by an interaction between 'improved' and 'common' farming methods.

Improved Practices

A very large number of farm implements and techniques were either introduced to or developed in Ireland during the last three centuries. The rate at which these came into widespread use is very difficult to assess, however. We have seen that the Royal Dublin Society was importing small numbers of new implements· even before 1750, and during the later eighteenth century some small-scale manufacturing of machines and tools was begun in several parts of Ireland. During the early nineteenth century all this activity seems to have accelerated. However, this is not necessarily evidence that the average Irish farmer was using the new technology. One interesting, although very tiny, survey carried out by the Royal Dublin Society in 1859 suggests that even among fairly affluent farmers adoption of new machinery was by no means common. Fourteen farmers from ten different counties were questioned about their farms, labour supply, and cultivation techniques. The farms varied considerably in size. The largest, in county Monaghan, had 1,964 acres under tillage, while one of the county Antrim farms had just over five acres of cultivated land. All of the farms made some use of horse-power, the number of working horses varying between one and six. Only two of the farms had reaping machines, which is not very surprising, since these had been introduced to Ireland just seven years before. However, horse-drawn hay-tedding machines were also found on only two of the farms, and only five had horse-powered threshing machines. (Returns for two of the other farms give prices for the hire of a steam thresher.)[4] The lack of use of these implements, which were proven to have great value in increasing both the possible scale and speed at which operations were performed, requires some explanation.

Despite the enthusiasm of agriculturalists, a generally cautious attitude to new technology was well justified. Only a very small proportion of the implements and techniques devised by improvers were in fact successful. Large capital inputs and the use of advanced engineering were no guarantees that experiments would lead to practical advances in technique. This can be clearly seen in attempts to apply steam power to ploughing. Steam ploughing equipment was advertised in *The Irish farmer's gazette* in 1863, and exhibited at the Royal Dublin Society's Spring Show in 1864.[5] In 1863 *Purdon's practical farmer* was enthusiastic: 'Seeing that not only the possibility, but the actually greater utility of steam cultivation has been sufficiently proved, we have not the slightest hesitation in saying that the sooner it becomes a regular part of our agricultural system the better'.[6] However, despite the continued manufacture of steam ploughing apparatus, it appears to have been very little used in Ireland, and by the early twentieth century, farming texts had become much more cautious in their assessment of steam power:

> As an adjunct to horse cultivation, steam is extremely valuable, and in many cases, it may partially displace horses. It is, however, noteworthy that after a trial of at least sixty years, it has hitherto failed to oust the horse from its position as the most generally employed power on the farm. . . The efficiency of steam cultivation does not. . . appear to be in all cases greater than that of horse tillage. Horse ploughing is quite as good as steam ploughing, and, as to ordinary harrowing, it is one of its greatest recommendations on many soils that the horses tread the ground and render it firm after sowing.[7]

Just how long, and how inconclusive, a debate between agricultural improvers could be is well illustrated by arguments about the relative efficiency of oxen and horses as draught animals. The systematic use of oxen in Ireland during the eighteenth and nineteenth centuries was almost always confined to large-scale farming. The level of interest in the potential of oxen, however, is suggested by the inclusion of 'Use of oxen – how harnessed', in the Royal Dublin Society's 'suggestions of enquiry' to compilers of the early nineteenth-century county statistical surveys. During this period, 'numberless statements'[8] were made on the subject, and discussion continued well into the present century.[9] At a 'theoretical' level the debate never reached a satisfactory conclusion, but at the level of practice the use of oxen dwindled away almost entirely. Some agriculturalists recognised that the debate did not reflect well on a purely theoretical approach to improvement. Oxen were said by one standard text to be preferred by 'gentlemen and theoretic farmers',[10] while another commented scathingly that, 'The preference has generally been given by speculative writers to the ox, and by practical farmers to the horse.'[11]

The differing and constantly changing views of theoretical agriculturalists were often criticised by contemporary observers as hindering rather than advancing the cause of improvement amongst farmers. E. Burroughs, commenting on the progress of agricultural improvement in Ireland, complained: 'We find so much difference of opinion, even among the most experienced writers, that the young practitioner is often puzzled to decide on which he is to place confidence'.[12] Not only did these diverse opinions confuse practical farmers, but more significantly they showed that the promotion of some improved practices and the rejection of some common practices by theorists were based on incomplete scientific knowledge. The *carte blanche* advocacy of trenching, which involved the inversion of the soil strata in deep tillage, as enhancing soil fertility, and the blanket rejection of paring and burning on the grounds of being an exhaustive process of land reclamation, aptly illustrate this lack of knowledge. Sproule sceptically remarked of early nineteenth-century attempts to apply scientific principles to agricultural practices: 'The art of agriculture has indeed been perfected by experiments and observation, rather than by deductions from principles of natural science; nor have we been yet able to found upon such deductions any very safe rules for the practice of husbandry'.[13]

The Persistence of 'Common' Practices

The reluctance of agriculturalists to admit that 'common' Irish cultivation implements and techniques had any merit whatsoever has been apparent throughout this book. Yet we have seen that within limited capital resources, and where labour was plentiful and cheap, 'common' practices could be efficient. Hand-tools might be simple in construction, but the techniques in which they were used were remarkable for refinement rather than crudity. Spadework and associated techniques of ridge cultivation show this very clearly, as does the use of sickles for harvesting grain. Indeed where there was so much under-employed labour, as was the case in early nineteenth-century Ireland, we have seen that some agriculturalists, as well as socially concerned contemporaries, advocated that labour-intensive methods, such as using flails for threshing, or spadework for large-scale tillage, should be used as much as possible. The low level of mechanisation of the farms in the Royal Dublin Society's survey mentioned earlier in this chapter was probably at least partly due to the availability of labour. All of the farms included had hired labour, one county Antrim farm employed 140 men and 120 women as extra hands during harvest.

The puzzlement of some agriculturalists when discussing common practices which still seemed to produce high yields is well illustrated by McParlan's comments on county Mayo in 1802:

> With the articles of implements I cannot have done without observing, that in general all through the county they are of the most inferior kind. The ploughs never sink deep enough, and the slight scratching they give the surface is left in unbroken lumps, by the light and wood-pinned harrows. . . the rakes, shovels, and all are bad. It has appeared, notwithstanding, how immense is the extent of tillage here, and the vast abundance exported of potatoes and corn.[14]

Even where implements had striking advantages in the speed and scale at which they could operate, or savings in draught they allowed, and where capital and labour were not limiting factors, a slow change to the new technology was by no means an indication of conservative prejudice. We have seen that 'common' Irish ploughs, while lacking the refined construction of new Scottish and English models, could be used efficiently in particular systems of cultivation. Early two-horse improved Scottish swing ploughs could only work to full advantage when ground was well drained, and where techniques such as the ploughing in of seed were not practised. Also, an infrastructure of blacksmiths and ploughmen well versed in the principles of the new, improved plough technology was necessary before the widespread manufacture and use of the implements was possible. This was lacking in late eighteenth- and early nineteenth-century Ireland, and as the development of such a tier of trained craftsmen and operatives takes time, it is not surprising that many farmers did not feel any urgent need to adopt the Scottish ploughs. New techniques, such as drilling grain seed, required a whole series of changes from tillage to harvest. The best agriculturalists recognised that a change in one aspect of a farming method could imply radical changes throughout the whole mode of cultivation.

Some integration of 'improved' and 'common' techniques was possible. Progressive, large-scale Irish farmers sometimes integrated elements from both systems, disregarding the more rigid strictures of theorists. One clear example of this was the use of high, narrow cultivation ridges in the initial stages of thorough drainage schemes, especially where marginal land was being reclaimed for cultivation. These reclamation schemes also led to the recognition that controlled paring and burning could accelerate the improvement of land. The integration of 'improved' and 'common' techniques was also practised by small farmers. The long-term modification of wooden swing ploughs in order to make potato drills in the Mourne area of county Down provides a good example. In county Kerry the combined use of swing plough, loy and shovel, with the adoption of the tractor for draught power in modern times, for lazy-bed construction, and in county Monaghan the use of chill plough and spade for six furrow ridges are others.

Perceptive agriculturalists were well aware that systems of farming developed according to abstract principles, or to suit the circumstances of farmers in lowland Scotland or southern England, would often require modification before they would work well in Irish conditions. Arthur Young recorded one disastrous attempt to introduce English methods to an estate in south-west Ireland. The landlord, wishing to improve an area of very rocky mountain, brought over two farmers from Sussex. He had houses and out-buildings erected for them, and let them have the land rent-free. The men sold up their holdings in England, and brought over their horses, stock and implements. The land, however, proved too stony to be cleared at anything other than very great expense. The project continued for four years, but ended with the landlord losing money, and the two farmers ruined. Young saw that such an imprudent undertaking 'was the sure way to bring. . . ridicule on what falsely acquired the name of *English husbandry'*.[15]

In 1812, Edward Wakefield, while criticising the construction and use of Irish farm implements, warned against the wholesale import of English implements as replacements:

> I have always found, that the mechanic of the district was best calculated to make [implements]. . . in such a manner as to suit the soil on which they are employed; for no one can give directions in regard to the form and shape of any instrument as well as the workman who uses it. . . A man who knows how to till the soil, wishes to have the tools adapted to his purposes; and, therefore, those constructed on general principles by a London mechanic, seldom answer so well as those under his own immediate inspection.[16]

The cultivation techniques described in this book have shown that Irish farmers cannot fairly be accused of stubborn intransigence. We can now re-examine some of the major inadequacies of Irish farming practice identified by contemporaries, and try and decide how far these defects resulted from deficiencies in farming at a technical level.

Minimal rotations and cropping to exhaustion were not inevitable results of Irish farming practice. The large proportion of arable land under grass, the frequent breaking up of pastures, and the natural fertility of soil in much of Ireland meant that good land, which remained largely under the control of commercial farmers, did not become exhausted.[17] The practice of 'bare' or 'summer' fallowing, where land was repeatedly ploughed during the year, also maintained the fertility of land. Agriculturalists criticised this type of fallowing, advocating the use of green cropping instead, but most criticisms were directed against the waste of land involved, rather than the effectiveness of the practice. (In 1802, Tighe claimed that summer fallowing could only be economically practised in county Kilkenny on farms of more than fifty acres.[18])

In areas of poorer land, where much of the expanding population in the late eighteenth and early nineteenth centuries was concentrated,[19] subdivision of holdings meant that the 'common' controls on soil deterioration could not be adhered to. It was in these areas that rotations were shortened to become a 'see-saw between oats and potatoes, with occasional rests',[20] and where paring and burning might be carried on until land was ruined. Hely Dutton's comment on the latter practice in county Clare shows the alternatives facing many small farmers: 'If a total abolition of this practice was to take place, as some people totally ignorant of rural economy seem to wish, a famine would be the consequence'.[21]

The claim that crops in Ireland were weed-infested can be related to the technique of broadcast-sowing cereals, and to late harvesting. Where crops were not sown in drills, weeding became extremely difficult, and the longer a crop was left standing, the more weeds grew amongst it. Reaping grain with sickles or hooks meant that weeds could be left behind during harvest, but how far weeds damaged crop yields remains uncertain. Any conclusions must be based on impressionistic evidence, since types of weeds, and their real extent, are not known. Agriculturalists differed in the emphasis they put on this problem. In 1835, it was alleged that a traveller in Ireland would see 'weeds often growing more luxuriant and numerous than even the grass or crops'.[22] However, this must be set against Arthur Young's comparison of the carefully weeded potato crops of the Palatines, in their colony at Adare, county Limerick, with those of surrounding farmers: '[The Palatines] are different from the Irish in several particulars; they put their potatoes in with the plough, in drills, horse-hoe them while growing, and plough them out. . . but the crops are not so large as in the common method'.[23] Careful husbandry was essential if the related set of drill implements were to be used, but Young, and the Irish, thought the Palatines went too far: 'Their industry goes so far, that jocular reports of its excess are spread: in a very pinching season, one of them yoked his wife against a horse, and went in that manner to work, and finished a journey at plough'.[24]

One reason why agriculturalists disapproved of Irish methods does seem to have been their untidy appearance. But this may have belied their effectiveness: a point which was recognised, not only by Arthur Young, but also by Hely Dutton who, as we have seen, warned landlords organising the drainage of land not to mistake 'neatness of execution for correctness of design'.[25] Horatio Townsend, discussing the practice of paring and burning in county Cork in 1810, suggested that, 'It is very probable that the naked appearance of land, let out without grass-seeds after burning, has been a principal cause of the objection to the mode. But this barrenness is more apparent than real'.[26] Another commentator, writing on the effects of cross ploughing during 'summer' or 'naked' fallowing, noted that the plough cut through the previously ploughed ridges in such a way that the furrows were 'generally thrown up in an irregular manner, and present to the inexperienced eye a confused aggregation of lumps of earth'. However, he claimed that despite the untidy appearance the technique was 'well adapted for exposing a very enlarged surface of soil to the ameliorating influence of the atmosphere', and served as a check on the growth of weeds.[27]

It would of course be absurd to present Irish farmers as uniformly good husbandmen. There are many practices recorded which it is difficult to regard as anything other than inefficient or slovenly since, even within 'common' practice, better alternatives were available. Probably the most notorious example of bad husbandry was ploughing by the tail, where draught horses' tails were substituted for trace ropes. Justifications of this practice – that it shortened the overall length of the plough team, or that it made horses more sensitive to obstructions met by the plough – are not convincing.[28] Alleged instances where the horses' tails were pulled off during ploughing would mean that, as harness, tails were not very reliable, and certainly very difficult to repair. We can label ploughing by the tail as bad husbandry because an efficient and widely used system of harnessing was within the resources of even small farmers. Collars made of plaited *súgán* (straw rope), and traces made from straw or withies, were recognised by contemporaries as easily made, cheap, and effective.[29]

In arguing that Irish farming methods were generally efficient, within the resources available, we do not want to suggest that agricultural research was in any way superfluous. The practical successes of this research have been as impressive as those of any other applied science. In 1863, *Purdon's practical farmer* admirably described the state of agricultural research:

We have yet much to learn, and. . . in general, we are only groping after, and have yet but faint conceptions of what are the true principles by which we should be guided, or the correct modes of reducing those principles to practice. But the knowledge of our deficiencies on these points ought to induce us to strain every nerve in order that all doubts may be cleared away, and that agriculture, the most important in its results of all the arts, shall be placed in its proper station, based upon and guided by the truths of general science.[30]

In any situation, however, there is a lag between the introduction of the discoveries arising from these worthy aims and their widespread adoption. Even in Norfolk, identified as one of the heartlands of the English agricultural revolution, the adoption of the famous four-course rotation was very slow.[31] Thirsk, discussing changes in English agricultural techniques, reaches a conclusion very similar to that suggested by Irish evidence: 'This survey of innovations underlines their hazards and uncertainties, which should not be brushed aside by after-knowledge. The risks of experiments were daunting to the great majority of farmers, who could not afford to gamble and lose'.[32] Contemporaries claimed that the lag between introduction and adoption, or at another level theory and practice, was wider in Ireland than in England and Scotland. We hope to have shown that at a technical level this lag did not imply inefficient farming. The history of flax, the most commercialised and least mechanised crop, shows very clearly how complex the relationship between technological change and economic development can be. The technical history of Irish agriculture is full of examples of the ingenuity with which farmers adopted or modified new techniques, if convinced of their value. Irish farmers were, however, very ready to ignore those which they did not regard as useful. As Solar has said, 'There were elements of the faith [Improvement], notably the use of improved implements and machinery, which were probably inappropriate to Irish labour supply conditions. All in all Irish agriculturalists were eclectics. They resisted repeated entreaties to conform and so earned the condemnation of the true believers'.[33] The success of their approach was demonstrated by the high yields produced by Irish commercial farming, and the skill with which the smallest subsistence farmers created a living out of previously waste land.

Examining the relationship between the resources available to small farmers, and their farming techniques, raises another point which is of central importance. This has been well made by Mokyr, who points out that although 'in a certain way. . . the Irish economy was well organised and [the Irish] were not throwing away resources. . . [This] is not quite the same as saying they were not poor'.[34] Farming techniques cannot be considered in isolation. The skills of ridgemaking and reaping developed by the Irish rural poor could not save vast numbers of them from extinction when the odds became overwhelming. We hope, however, to have shown that when examining the farming techniques used by these people, we should not wonder why they did not improve, but instead express full appreciation of the ingenious methods which enabled so many of them to survive for so long.

NOTES

1. Solar, Peter, 'Agricultural productivity and economic development in Ireland and Scotland in the early nineteenth century', in *Ireland and Scotland 1600-1850*, eds. T.M. Devine and D. Dickson (Edinburgh, 1983), p.76.

2. Tighe, W., *Statistical observations relative to the county of Kilkenny* (Dublin, 1802), p.178.

3. Loudon, J.C., *An encyclopaedia of agriculture* (London, 1831), p.135.

4. Hamilton, Charles W., 'Abstract of answers to the Royal Dublin Society's agricultural queries of 1859', Royal Dublin Society, *Journal*, vol. 3 (Dublin, 1860), pp.24-25.

5. *The Irish farmer's gazette*, vol. 22 (Dublin, 1863), p.483; *ibid.*, vol. 23 (Dublin, 1864), p.135.

6. *Purdon's practical farmer* (Dublin, 1863), pp.198-199.

7. Wright, R.P. (ed.), *The standard cyclopaedia of modern agriculture and rural economy*, vol. II (London, 1908), p.112.

8. Library of Useful Knowledge, *British husbandry*, vol. 1 (London, 1834), p.180.

9. Russell, George (A.E.) (ed.), *The Irish homestead* (Dublin, 1917), pp.189-190.

10. Library of Useful Knowledge, *op. cit.*, vol. 1, p.180.

11. Loudon, J.C., *op. cit.*, p.782.

12. Burroughs, Edward, *The Irish farmer's calendar* (Dublin, 1835), p.iii.

13. Sproule, John, *A treatise on agriculture suited to the soil and climate of Ireland* (Dublin, 1839), p.iv.

14. McParlan, James, *Statistical survey of the county of Mayo* (Dublin, 1802), p.37.

15. Young, Arthur, *A tour in Ireland*, vol. 2 (Dublin, 1780), pp.81-82.

16. Wakefield, E., *An account of Ireland, statistical and political*, vol. 1 (London, 1812), p.506.

17. Solar, Peter, *op. cit.*, pp.76-77.

18. Tighe, W., *op. cit.*, p.410.

19. Solar, Peter, *op. cit.*, p.77.

20. Micks, W.L., *An account of the Congested Districts Board for Ireland from 1891 to 1923* (Dublin, 1925), p.243.

21. Dutton, Hely, *Statistical survey of the county of Clare* (Dublin, 1808), p.37.

22. Anon., 'On agricultural schools', *The Irish farmer's and gardener's magazine*, vol. 2 (Dublin, 1835), pp.402-403.

23. Young, Arthur, *op. cit.*, vol. 2, p.138.

24. *Ibid.*, pp.138-139.

25. Dutton, Hely, *Statistical and agricultural survey of the county of Galway* (Dublin, 1824), p.170.

26. Townsend, H., *Statistical survey of the county of Cork* (Dublin, 1810), p.544.

27. Sproule, John, *op. cit.*, pp.171-172.

28. Lucas, A.T., 'Irish ploughing practices, part two', *Tools and tillage*, vol. 2:2 (Copenhagen, 1973), pp.72-79.

29. *Ibid.*, pp.195-204.

30. Purdon, *op. cit.*, pp.1-2.

31. Grigg, David, *The dynamics of agricultural change* (London, 1982), p.182.

32. Thirsk, Joan, 'Agricultural innovations and their diffusion', in *The agrarian history of England and Wales*, vol. 5:2, ed. Joan Thirsk (Cambridge, 1985), p.587.

33. Solar, Peter, *op. cit.*, pp.80-81.

34. Devine, T.M. and Dickson, David, 'A review of the symposium', in *Ireland and Scotland 1600-1800*, eds. T.M. Devine and D. Dickson (Edinburgh, 1983), p.269.

Bibliography

Adams, G.B., 'Work and words for haymaking', *Ulster Folklife*, vol. 12 (Holywood, 1966).

Andrews, J.H., 'Limits of agricultural settlement in Pre-Famine Ireland', in *Ireland and France, 17th-20th centuries*, eds. L.M. Cullen and F. Furet (Paris: École des Hautes Études, 1980).

Anon., 'On agricultural schools', *The Irish farmer's and gardener's magazine*, vol. 2 (Dublin, 1835).

Archer, Joseph, *Statistical survey of the county Dublin* (Dublin, 1801).

The Ark (Belfast, 1910).

Baker, John Wynn, *Experiments in agriculture, made under the direction of the Right Honourable Dublin Society in the year 1767* (Dublin, 1769).

Baker, John Wynn, *A short description and list, with the prices of the instruments of husbandry, made in the factory at Laughlinstown, near Cellbridge, in the country of Kildare* (Dublin, 1767-1769).

Baldwin, Thomas, *Handy book of small farm management* (Dublin, 1870).

Baldwin, Thomas, *Introduction to Irish farming* (London, 1874).

Baldwin, Thomas, *Introduction to practical farming* (Dublin, 1877).

Baldwin, Thomas, *Introduction to practical farming*, 23rd ed. (Dublin, 1893).

Beaufort, D.A., 'Materials for the Dublin Society agricultural survey of county Louth', ed. C.C. Ellison, *Journal of the County Louth Archaeological and Historical Society*, vol. 18:1 (Dundalk, 1973).

Bede, *A history of the English church and people*, trans. Leo Sherley-Price, rev. ed. R.E. Latham (Harmondsworth: Penguin, 1983).

Belfast Commercial Chronicle (Belfast, 1st October, 1806).

Belfast Newsletter (Belfast, 3rd August, 1750).

Bell, J., 'Wooden ploughs from the mountains of Mourne, Ireland', *Tools and tillage*, vol. 4:1 (Copenhagen, 1980).

240

Bell, J., 'Sickles, hooks and scythes in Ireland', *Folklife*, vol. 19 (Leeds, 1981).

Bell. J., 'Harrows used in Ireland', *Tools and tillage*, vol. 4:4 (Copenhagen 1983).

Bell, J., 'The use of oxen on Irish farms since the eighteenth century', *Ulster Folklife*, vol. 29 (Holywood, 1983).

Bell, J., 'A contribution to the study of cultivation ridges in Ireland', *Journal of the Royal Society of Antiquaries of Ireland*, vol. 114 (Dublin, 1984).

Berry, Henry F., *A history of the Royal Dublin Society* (London, 1915).

Binchy, D.A. (ed.), *Críth gablach* (Dublin: Stationery Office, 1941).

Binns, J., *Miseries and beauties of Ireland*, 2 vols. (London, 1837).

Board of Agriculture (England), 'Haymaking' Leaflet no. 85, Department of Agriculture and Technical Instruction for Ireland *Journal*, vol. 3 (Dublin: H.M.S.O., 1903).

Buchanan, R.H., 'Calendar customs, part 1', *Ulster Folklife*, vol. 8 (Belfast, 1962).

Burroughs, E., *The Irish farmer's calendar* (Dublin, 1835).

Cassell's household guide, vol. 4 (London, 18?).

Caulfield, Seamus, 'Neolithic fields: the Irish evidence' (British Archaeological Report, 48), eds. H. Bowen and P. Fowler (Oxford, 1978).

Clark, A.L., 'Historical sketch of the flax-growing industry', Department of Agriculture and Technical Instruction for Ireland *Journal*, vol. 3 (Dublin: H.M.S.O., 1903).

Clark, A.L., 'Flax seed for sowing purposes', Department of Agriculture and Technical Instruction for Ireland *Journal*, vol. 4 (Dublin: H.M.S.O., 1904).

Dubordieu, John, *Statistical survey of the county of Down* (Dublin, 1802).

Dubordieu, John, *Statistical survey of the county of Antrim* (Dublin, 1812).

Duignan, M., 'Irish agriculture in early historic times', *Journal of the Royal Society of Antiquaries of Ireland*, vol. 74 (Dublin, 1944).

The Dungannon news (Dungannon, 1st May, 1902).

Dutton, Hely, *Observations on Mr. Archer's statistical survey of the county of Dublin* (Dublin, 1802).

Dutton, Hely, *Statistical survey of the county of Clare* (Dublin, 1808).

Dutton, Hely, *A statistical and agricultural survey of the county of Galway* (Dublin, 1824).

Evans, E.E., *Irish heritage* (Dundalk, 1949).

Evans, E.E., *Irish folkways* (London, 1957 (1967)).

'F', 'Observations on the agriculture of the north of Ireland', *The Irish farmer's and gardener's magazine*, vol. 1 (Dublin, 1834).

The farmer's friend (London, 1847).

The farmer's gazette, vols. 7-9 (Dublin, 1847-1849).

Fenton, A., *Scottish country life* (Edinburgh, 1976).

Fussell, G.E., *The farmer's tools, 1500-1900* (London, 1952).

Gailey, R.A., 'Irish iron-shod wooden spades', *Ulster journal of archaeology*, vol. 31 (Belfast, 1968).

Gailey, R.A., 'The typology of the Irish spade', in *The spade in northern and Atlantic Europe*, eds. R.A. Gailey and A. Fenton (Belfast: Ulster Folk Museum, 1970).

Gailey, R.A., 'Spade tillage in south-west Ulster and north Connaught', *Tools and tillage*, vol. 1:4 (Copenhagen, 1971).

Gailey, R.A., *Spade making in Ireland* (Holywood: Ulster Folk and Transport Museum, 1982).

Gailey, R.A., 'Introduction and spread of the horse-powered threshing machine to Ulster's farms in the nineteenth century: some aspects', *Ulster Folklife*, vol. 30 (Holywood, 1984).

Greig, W., *General report on the Gosford estates in county Armagh*, introd. F.M.L. Thompson and D. Tierney (Belfast: H.M.S.O., 1976).

Grigg, David, *The dynamics of agricultural change* (London, 1982).

Hale, Thomas, *A compleat body of husbandry*, 4 vols. (Dublin, 1757).

Hall, Mr. and Mrs. S.C., *Ireland: its scenery, character etc.*, vol. 1 (London, c.1850).

Hamilton, Charles W., 'Abstract of answers to the Royal Dublin Society's agricultural queries of 1859', Royal Dublin Society *Journal*, vol. 3 (Dublin, 1860).

Harkin, Maura and McCarroll, Sheila, *Carndonagh* (Dublin, 1984).

Hill, George, *Facts from Gweedore*, introd. E.E. Evans (Belfast: Queen's University, Institute of Irish studies, 1971).

Howatson, W., 'Grain harvesting and harvesters', in *Farm servants and labourers in lowland Scotland*, ed. T.M. Devine (Edinburgh, 1984).

Hyndman, C., *A new method of raising flax* (Belfast, 1774).

Impartial reporter (Enniskillen, 3rd August, 1967).

Instructions to owners applying for loans for farm buildings (Dublin, 1856).

The Irish farmer's and gardener's magazine, vol. 1 (Dublin, 1834).

Collins, E.J.T., *Sickle to combine* (Reading: Museum of English Rural Life, 1969).

Congested Districts Board for Ireland, *Twelfth annual report* (Dublin: H.M.S.O., 1903).

Connell, K.H., 'The colonization of waste land in Ireland, 1780-1845', *Economic history review*, vol. 3:1 (London, 1950).

Coote, C., *Statistical survey of the county of Cavan* (Dublin, 1802).

Coote, C., *Statistical survey of the county of Armagh* (Dublin, 1804).

Cotter, Rev. G.S., 'Advice to the farmers of the county of Cork, on the culture of wheat', *The Munster farmer's magazine*, vol. 1 (Cork, 1812).

County agricultural instructors, *Agriculture in Ireland* (Dublin: *The farmer's gazette*, 1907).

Crawford, W.H., *Economy and society in eighteenth century Ulster* Ph.D. thesis (Belfast: Queen's University, 1982).

Cullen, L.M., *Life in Ireland* (London, 1968).

Cullen, L.M., *An economic history of Ireland since 1600* (London, 1972).

Curran, Simon, 'The Society's role in agriculture since 1800', in *The Royal Dublin Society, 1731-1981*, eds. J. Meenan and D. Clarke (Dublin, 1981).

Davies, C., 'Kilns for flax-drying and lime-burning', *Ulster journal of archaeology*, 3rd series, vol. 1 (Belfast, 1938).

Davis, Francis, 'Low and clean', in *Earlier and later leaves* (Belfast, 1878).

De Latocnaye, *A Frenchman's walk through Ireland, 1796-7*, trans. J. Stevenson (Belfast, 1917).

Dennchy, Mary, 'Agricultural education in the nineteenth century', *Retrospect: Journal of the Irish history students' association*, vol. 5 (Dublin, 1982).

Department of Agriculture and Technical Instruction for Ireland, 'Flax experiments, 1901', *Journal*, vol. 2 (Dublin: H.M.S.O., 1902).

Department of Agriculture and Technical Instruction for Ireland, 'Prevention of potato blight', *Journal*, vol. 2 (Dublin: H.M.S.O., 1902).

Department of Agriculture and Technical Instruction for Ireland, 'Statistical survey of Irish agriculture', *Ireland, industrial and agricultural* (Dublin, 1902).

Department of Agriculture and Technical Instruction for Ireland, 'Haymaking', Leaflet no. 46, *Journal*, vol. 4 (Dublin: H.M.S.O., 1904).

Department of Agriculture and Technical Instruction for Ireland, *Twelfth annual report, 1911-1912* (London: H.M.S.O., 1913).

Devine, T.M. and Dickson, David, 'A review of the symposium', in *Ireland and Scotland 1600-1800*, eds. T.M. Devine and David Dickson (Edinburgh, 1983).

[Devon Commission], *Digest of evidence taken before Her Majesty's Commissioners of Inquiry into the state of the law and practice in respect to the occupation of land in Ireland*, 2 parts (Dublin, 1847).

Dineen, Patrick S., *Foclóir Gaedhilge agus Béarla* (Dublin: Irish Texts Society, 1927).

Doyle, Martin, *Hints originally intended for the small farmers of the county of Wexford* (Dublin, 1830).

Doyle, Martin, *Hints to smallholders on planting and on cattle etc.* (Dublin, 1830).

Doyle, Martin, 'On agricultural schools', *The Irish farmer's and gardener's magazine*, vol. 1 (Dublin, 1834).

Doyle, Martin, *A cyclopaedia of practical husbandry*, rev. ed. W. Rham (London, 1844).

Dublin penny journal, vols. 1-2 (Dublin, 1833-1834).

Dubois, Edward, *My pocket book; or hints for "a ryghte merrie and conceitede" tour* (London, 1808).

The Irish farmer's gazette (Dublin, 1863-1864).

Irish National Schools, *Agricultural classbook* (Dublin: Commissioners of National Education, 1868).

Irish National Schools, *Introduction to practical farming*, 2 vols. (Dublin, 1898).

Joyce, P.W., *A social history of ancient Ireland*, 2 vols. (Dublin, 1920).

Kennedy, David, 'Templemoyle agricultural seminary, 1827-1866', *Studies: an Irish quarterly review*, vol. 29 (Dublin, 1940).

Lambert, Joseph, *Observations on the rural affairs of Ireland* (Dublin: Curry, 1829).

Lambert, Joseph, *Agricultural suggestions to the proprietors and peasantry of Ireland* (Dublin: *Farmer's gazette*, 1845).

Lane, Padraig, *Ireland* (London, 1974).

Library of Useful Knowledge, *British husbandry*, 2 vols. (London, 1834).

Loudon, J.C., *An encyclopaedia of agriculture* (London, 1831).

Low, D., *Elements of practical agriculture* (Edinburgh, 1834).

Lucas, A.T., 'Making wooden sieves', *Journal of the Royal Society of Antiquaries of Ireland*, vol. 81:2 (Dublin, 1951).

Lucas, A.T., 'An fhóir: a straw rope granary', *Gwerin*, vol. 1 (Oxford, 1957).

Lucas, A.T., 'An fhóir: a straw rope granary' *Gwerin*, vol. 2 (Oxford, 1958).

Lucas, A.T., 'Paring and burning in Ireland; a preliminary survey', in *The spade in northern and Atlantic Europe*, eds. R.A. Gailey and A. Fenton (Belfast: Ulster Folk Museum, 1970).

Lucas, A.T., 'The "gowl-gob", an extinct spade type from county Mayo, Ireland', *Tools and tillage*, vol. 3:4 (Copenhagen, 1979).

Lucas, A.T., 'Irish ploughing practices', 4 parts, *Tools and tillage*, vol 2:1-4 (Copenhagen, 1972-1975).

McEvoy, John, *Statistical survey of the county of Tyrone* (Dublin, 1802).

MacGill, Patrick, *Children of the dead end* (London, 1914).

MacNeill, M., *Lughnasa* (London, 1962).

McParlan, James, *Statistical survey of the county of Donegal* (Dublin, 1802).

McParlan, James, *Statistical survey of the county of Leitrim* (Dublin, 1802).

McParlan, James, *Statistical survey of the county of Mayo* (Dublin, 1802).

McParlan, James, *Statistical survey of the county of Sligo* (Dublin, 1802).

McCully, James, *Letters by a farmer* (Belfast, 1787).

Marshall, G., 'The "Rotherham" plough', *Tools and tillage* vol. 4:3 (Copenhagen, 1982).

Mason, Thomas H., *The islands of Ireland*, 2nd ed. (London, 1938).

Mason, William Shaw, *A statistical account, or parochial survey of Ireland*, 3 vols. (Dublin, 1814).

Meenan, J. and Clare, D., 'The Royal Dublin Society', in *The Royal Dublin Society, 1731-1981*, eds. J. Meenan and D. Clarke (Dublin, 1981).

Micks, W.L., *An account of the. . . Congested Districts Board for Ireland from 1891 to 1923* (Dublin, 1925).

Mitchell, Frank, *The Irish landscape* (London, 1976).

Mogey, J.M., *Rural life in northern Ireland* (London, 1947).

Mokyr, Joel, *Why Ireland starved* (London, 1983).

Mullen, Pat, *Man of Aran* (London, 1934).

The Munster farmer's magazine (Cork, 1812-1817).

Murphy, Edmund, 'Agricultural report', *The Irish farmer's and gardener's magazine*, vol. 1 (Dublin, 1834).

Murphy, Edmund, 'Observations made on a visit to the agricultural seminary at Templemoyle, in the county of Londonderry', *The Irish farmer's and gardener's magazine*, vol. 1 (Dublin, 1834).

Murphy, Edmund, 'Spade husbandry', *The Irish farmer's and gardener's magazine,* vol. 1 (Dublin, 1834).

Murphy, Edmund, *The Irish farmer's and gardener's register,* vol. 1 (Dublin, 1863).

Murphy, M.J., 'The mower from Moygannon', *Ulster Folklife,* vol. 12 (Holywood, 1966).

Nicholson, Asanath, *The Bible in Ireland.* Quoted in Frank O'Connor (ed.), *A book of Ireland* (Glasgow, 1959 (1974)).

North West of Ireland Society *Magazine,* vol. 1 (Derry, 1823).

Northern Whig (Belfast, 7th May, 1842).

Ó Danachair, Caiomhín, 'The flail and other threshing methods', *Journal of the Cork Historical and Archaeological Society,* vol. 60:19 (Cork, 1955).

Ó Danachair, Caoimhín, 'The spade in Ireland', *Béaloideas,* vol. 31 (Dublin, 1963).

Ó Danachair, Caiomhín, 'The use of the spade in Ireland', in *The spade in northern and Atlantic Europe,* eds. R.A. Gailey and A. Fenton (Belfast: Ulster Folk Museum, 1970).

Ó Danachair, Caiomhín, 'The flail in Ireland', *Ethnologia Europaea,* vol. 4 (Arnhem, 1971).

Ó Donaill, N., *Foclóir Gaeilge-Béarla* (Dublin: Oifig an tSoláthair, 1977).

O'Dowd, Anne, *Meitheal: a study of co-operative labour in rural Ireland* (Dublin: Comhairle Bhéaloideas Éireann, 1981).

O'Loan, John, 'The manor of Cloncurry, Co. Kildare, and the feudal system of land tenure in Ireland', Department of Agriculture (Ireland) *Journal,* vol. 58 (Dublin, 1961).

O'Loan, John, 'A history of early Irish farming', 3 parts, Department of Agriculture (Ireland) *Journal, vols.* 60-62 (Dublin, 1963-1965).

O'Neill, Thomas P., 'The scientific investigation of the failure of the potato crop in Ireland, 1845-6', *Irish historical studies,* vol. 5:18 (Dublin, 1846).

O'Neill, Timothy P., *Life and tradition in rural Ireland* (London: Dent, 1977).

O'Rahilly, T.F., 'Etymological notes, 2', *Scottish Gaelic studies*, vol. 2 (Edinburgh, 1927).

Ordnance survey memoirs Mss, Royal Irish Academy, Dublin.

Ó Súilleabháin, Amhlaiobh, *Cinnlae Amhlaiobh Úi Shúilleabháin*, vol. 3, trans. M. McGrath (London: Irish Texts Society, 1930 (1936)).

Ó Súilleabháin, Michael, 'The bodhran, parts 1 and 2', *Treoir*, vol. 6 (Dublin, 1974).

'Parliament chloinne Tomáis', trans. O.J. Bergin, *Gadelica* (Dublin, 1912).

Parry, M.L., 'A typology of cultivation ridges in southern Scotland', *Tools and tillage*, vol. 3:1 (Copenhagen, 1976).

Partridge, Michael, *Farm tools through the ages* (Reading, 1973).

Patents for inventions: abridgments of specifications, class 6, agricultural appliances. . . A.D. 1855-1856 (London: H.M.S.O., 1905).

Perkins, J.A., 'Harvest technology and labour supply in Lincolnshire and the East Riding of Yorkshire, part 2', *Tools and tillage*, vol. 3:2 (Copenhagen, 1977).

Pierce Ltd., *The story of Pierce* (Wexford, 197?).

Purdon's practical farmer (Dublin, 1863).

Rawson, T.J., *Statistical survey of the county of Kildare* (Dublin, 1807).

Regulations for gathering seaweed, Ardglass, Co. Down, Public Records Office of Northern Ireland. Document T1009/160.

Roberts, Michael, 'Sickles and scythes: women's work and men's work at harvest time', *History workshop*, vol. 7 (Southampton, 1979).

Russell, George (A.E.) (ed.), *The Irish homestead* (Dublin, 1917).

Salaman, R.N., 'The influence of the potato on the course of Irish history' (The Tenth Finlay Memorial Lecture) (Dublin, 1944).

Salaman, R.N., *The history and social influence of the potato* (Cambridge, 1949 (1970)).

Sampson, G.V., *A memoir, explanatory of the chart and survey of the county of London-derry, Ireland* (London, 1814).

Silcock, James *et al*, 'The influence of rippling', Department of agriculture and technical instruction for Ireland *Journal,* vol. 3 (Dublin: H.M.S.O., 1903).

Smith, James, *Remarks on thorough draining and deep ploughing* (Stirling, 1843).

Solar, Peter, 'Agricultural productivity and economic development in Ireland and Scotland in the early nineteenth century', *Ireland and Scotland, 1600-1850,* eds. T.M. Devine and D. Dickson (Edinburgh, 1983).

Sproule, John, *A treatise on agriculture* (Dublin, 1839).

Sproule, John (ed.), *The Irish industrial exhibition of 1853: a detailed catalogue of its contents* (Dublin, 1854).

Stephens, Henry, *Book of the farm,* 2 vols., 3rd ed. (Edinburgh, 1871).

Stephens' book of the farm, 3 vols., 5th ed. rev. J. Macdonald (Edinburgh, 1908).

Thackeray, William Makepeace, *The Irish sketch-book of 1842 (Works,* vol. 18) (London, 1879).

Thirsk, Joan, 'Agricultural innovations and their diffusion', in *The agrarian history of England and Wales,* vol. 5:2 ed. Joan Thirsk (Cambridge, 1985).

Thompson, R., *Statistical survey of the county of Meath* (Dublin, 1802).

Tighe, W., *Statistical observations relative to the county of Kilkenny* (Dublin, 1802).

Townsend, H., *Statistical survey of the county of Cork* (Dublin, 1810).

Trench, W.S., *Realities of Irish life* (London, 1868).

Ulster farmer and mechanic (Belfast, 1824-25).

Vaughan, W.E., 'Landlord and tenant relations in Ireland between the famine and the land war, 1850-78', in *Comparative aspects of Scottish and Irish economic and social history,* eds. L.M. Cullen and T.C. Smout (Edinburgh, 1978).

Wakefield, E., *An account of Ireland, statistical and political,* 2 vols. (Dublin, 1812).

Wallace, M.G., 'Early potato growing', Department of Agriculture and Technical Instruction for Ireland *Journal,* vol. 6 (Dublin: H.M.S.O., 1906).

Watson, M., 'Flachters: their construction and use on an Ulster peat bog', *Ulster Folklife*, vol. 25 (Holywood, 1979).

Watson, M., 'Cushendall hill ponies', *Ulster Folklife*, vol. 26 (Holywood, 1980).

Watson, M., 'North Antrim swing ploughs: their construction and use', *Ulster Folklife*, vol. 28 (Holywood, 1982).

Watson, M., 'Common Irish plough types and tillage practices', *Tools and tillage*, vol. 5:2 (Copenhagen, 1985).

The weekly agricultural review, Dublin, 15th April, 1859.

Weld, Isaac, *Statistical survey of the county of Roscommon* (Dublin, 1832).

Wexford Engineering Company, *Catalogue* (Wexford, 1950).

Wilde, W.R., *Catalogue of the antiquities of the museum of the Royal Irish Academy* (Dublin, 1857).

Wright, R.P. (ed.), *The standard cyclopaedia of modern agriculture and rural economy*, 12 vols. (London, 1909).

Young, Arthur, *A tour in Ireland*, 2 vols. (Dublin, 1780).

Young, R.M., *Historical notes of old Belfast* (Belfast, 1896).

Tapes

C77.49	Interview with Mr. William Donnelly, Lisburn, Co. Antrim.
C77.116	Interview with Mr. J.C. Drennan, Limavady, Co. Derry
C78.54	Interview with Mr. J.A. Weir, Ballyroney, Co. Down.
R79.39	Interview with Mr. Patrick Brogan, Hornhead, Co. Donegal.
R80.39	Interview with Mr. Cormac McFadden, Roshin, Co. Donegal.
R82.73	Interview with Mr. Archie Mullen, Armoy, Co. Antrim.
R82.95	Interview with Richard, Eugene and Felix McConville, Dromore, Co. Down.
R84.40	Interview with Mr. Joe Kane, Drumkeeran, Co. Fermanagh.
R85.143	Interview with Mr. John Joe McIlroy, Derrylea, Co. Monaghan.

Index